# WMD Terrorism

## Strategy for Consequence Management

# WMD Terrorism

## Strategy for Consequence Management

Brig MS Patial

(Established : 1870)

United Service Institution of India
Prof. D S Kothari Chair

Manas Publications
New Delhi-110002 (INDIA)

**Manas Publications**
(Publishers, Distributors, Importers & Exporters)
4858, Prahlad Street,
24, Ansari Road, Darya Ganj,
New Delhi - 110 002 (INDIA)
**Ph.: 23260783, 23265523:, 23842660 (R)**
**Fax: 011-23272766**
E-mail: manaspublications@vsnl.com



ISBN 81-7049-253-X
Rs. 495/-



The views expressed in this book are those of the author and not necessarily of the United Service Institution of India or the publisher. The USI or the publisher are not responsible for the views of the author and authenticity of the data, in any way what so ever.

*Typeset at*
Manas Publications

*Printed in India at*
Salasar Imaging Systems, Delhi
and Published by Mrs Suman Lata for
**Manas Publications**, 4858, Prahlad Street,
24, Ansari Road, Darya Ganj,
New Delhi - 110 002 (INDIA)

# Contents

# Acknowledgements

I am grateful to the United Service Institution of India (USI) for having given me the opportunity to undertake the fellowship to carry out research on such a important subject as Strategy for Consequence Management of Weapons of Mass Destruction (WMD) Terrorism. This fellowship provided me an opportunity to interact freely with both serving and retired officers from the Armed Forces and Professors and academicians from Institutions of repute.

My special thanks to Lieutenant General Satish Nambiar, PVSM, AVSM, VrC (Retd), Director, United Service Institution of India and Major General YK Gera (Retd), USI's Deputy Director and Editor, for providing me an ideal research environment and encouraging me from time to time to achieve high standards. I am extremely grateful to my colleagues Brigadier Balbir Pama, Colonel SC Tyagi and Lieutenant Colonel TV Rao for giving me rational and balanced advice and guidance. My thanks to Mrs. Kiran for typing the manuscript in its various stages.

For the last twenty years I have had the opportunity of not only reading about terrorism, but also physically experiencing the phenomenon of terrorism. While serving in Jammu and Kashmir, I have been involved in interrogating terrorists both, of local and foreign origin. I also had an opportunity to be the first among others to raise and serve in the elite National Security Guards, the only organisation in our country that is trained and equipped to carry out counter-terrorism and anti-terrorism operations against terrorists. I have also attended Workshop on Weapons of Mass Destruction (WMD) Consequence Management at Washington DC in October 2002. In my capacity as Director (NBCW), in the Army Headquarters, I have had many occasions to attend and participate in numerous lectures, workshops and seminars on the subject, from where I have collected a wealth of material. Most of the material used in this study is from such sources.

Also in this complete research process, I have referred to many primary original source material and open source materials, which I have quoted, where applicable. In addition I have interacted with a number of persons with whom I have met and discussed and whose names cannot be revealed for security reasons. Reference to many periodicals and newspapers may have been done over and above my personal impressions, all of which cannot be quoted, yet I acknowledge their immense value regarding information provided.

Lastly I am extremely grateful to my wife Madhu, son Anshuman and daughter Ayesha who encouraged me throughout the period of this research and gave a number of valuable suggestions regarding the language and layout of the book.

# Preface

Neither terrorism nor the proliferation of weapons of mass destruction (WMD) are new phenomena; states in key regions of the world's security concern have for several years aggressively pursued nuclear, biological or chemical weapons and missile capabilities or have engaged in or sponsored terrorism. What is new is the prospective conjuncture of these twin scourges that constitutes a combined threat. Terrorism has not only redefined the 'rules of engagement' but also how wars will be fought in the future. Globalisation, which was being hailed as a catalyst for shrinking distances and barriers, is suddenly being looked at with suspicion. A major reason for terrorism's increased lethality is the active role played by some states in supporting terrorism, resulting in 'force multiplying' effect. It places greater resources at the disposal of the terrorists, thereby enhancing planning, intelligence, logistical capabilities, training, finances and sophistication. The capability of terrorists groups to cause mass casualities and psychological damage through the use of weapons of mass destruction is a cause for international concern.

Both the external and internal environment of states are changing at an incredibly fast pace, with increasing cross border terrorism, the emergence of non-state actors, the growth of Islamic fundamentalism, the narcotics-arms nexus, illegal migration and left wing extremism, gravely impacting upon the security of the country. The rapid technological developments underway at the same time not only facilitate these events by reducing our reaction time but also add entirely new dimensions of threats and challenges. International and state sponsored terrorism, often motivated by fundamentalist ideologies, backed by secretive but efficient financial networks, use of IT, clandestine access to chemical, biological and nuclear materials and illicit drug trafficking, has emerged as a major threat to international stability and poses a challenge to the multi-religious, multi-ethnic and pluralistic composition of societies. India, thus, is one of the countries at the receiving end of these violent elements.

To deter this threat would be the prime objective but preparedness to fight the after-effects of an attack are equally important. It is for this reason that a comprehensive national strategy to deal with this threat is needed. The best defence against weapons of mass destruction and terrorism is good offence, backed up by effective deterrent, operational and mitigative plans and capabilities. India needs to pursue a proactive policy against the security challenges emerging from WMD-terrorism nexus and strengthen and transform its military and civilian response capability to deter, protect against and mitigate the effects of such threats. It will have to bolster its NBC detection capabilities and work out a comprehensive plan to mitigate the consequences of WMD terrorism. Adherence to UN norms and international cooperation with other nation states to deal with weapons of mass destruction terrorism will pay rich dividends.

An in-depth analysis has been done on issues of terrorism, weapons of mass destruction and their likely use by terrorists. The strategy India needs to adopt for consequence management of WMD terrorism has been addressed holistically.

**Brig MS Patial**

# Acronyms

| | |
|---|---|
| ADM | Atomic Demolition Munitions |
| AT | Anti Terrorist |
| BTWC | Biological and Toxin Weapons Convention |
| BW | Biological Warfare |
| CBIRF | Chemical Biological Incident Response Force |
| CISF | Central Industrial Security Force |
| CTBT | Comprehensive Nuclear Test Ban Treaty |
| CTR | Co-operative Threat Reduction |
| CW | Chemical Warfare |
| CWC | Chemical Weapons Convention |
| DoD | Department of Defense |
| EMP | Electromagnetic Impulse |
| EMS | Emergency Medical Services |
| EOD | Equipment Ordnance Disposal |
| EU | European Union |
| FBI | Federal Bureau of Investigation |

| | |
|---|---|
| FEMA | Federal Emergency Management Agency |
| GZ` | Ground Zero |
| HAZMAT | Hazardous Material |
| HYCE | High Yield Conventional Explosive |
| HEU | Highly Enriched Uranium |
| IAEA | International Atomic Energy Agency |
| IIZ | Initial Isolation Zone |
| IC | Incident Commander |
| ICS | Incident Command System |
| IED | Improvised Explosive Device |
| IND | Improvised Nuclear Device |
| IPE | Individual Protection Equipment |
| ITBP | Indo Tibetan Border Police |
| LFA | Lead Federal Agency |
| MCI | Mass Casualty Incident |
| MTCR | Missile Technology Control Regime |
| NBC | Nuclear, Biological and Chemical Weapons |
| NDMA | National Disaster Management Authority |
| NEMA | National Emergency Management Agency |
| NGO | Non-Governmental Organisation |
| OPCW | Organisation for the Prohibition of Chemical Weapons |
| PAZ | Protective Action Zone |
| PNE | Peaceful Nuclear Explosion |
| RDD | Radiological Dispersion Device |

## Acronyms

| | |
|---|---|
| SARS | Severe Acute Respiratory Syndrome |
| SEMA | State Emergency Management Agency |
| SOP | Standard Operating Procedure |
| TIC | Toxic Industrial Chemicals |
| TREE | Transient Radiation Effects on Electronics |
| UN | United Nations |
| WMD | Weapons of Mass Destruction |

# 1

## Introduction

*I know not with what weapons World War III will be fought, but World War IV will be fought with sticks and stones.*

**- Albert Einstein**

Terrorism has been with mankind for centuries and it has always attracted inordinate attention because of its dramatic character and its sudden, often wholly unexpected occurrence. It has been a tragedy for the victims, but seen in historical perspective it seldom has been more than a nuisance. Even the bloodiest terrorist incidents in the past have affected relatively few people. Yesterday's nuisance has become one of the gravest dangers facing mankind today. For the first time in history, weapons of enormous destructive power are both readily acquired and harder to track. In this new age, even the cost of hundreds of lives may appear small in retrospect. Science and technology has made enormous progress but human nature has not changed. There is as much fanaticism and madness as there ever was and there are now very powerful weapons of mass destruction available to terrorists.

Presently it is technologically possible to kill numerous people, not to mention the toll panic that is likely to ensue may take. There has been a radical transformation, if not a revolution, in the character of terrorism. It is not suggested that most terrorist groups will use weapons of mass destruction (WMD) in the near future; most of them will probably not. The problem of their accessibility and the technical difficulties in the way of effective use of arms of WMD are considerable. But the danger is so grave, the consequences so incalculable, that even the occurrence of few such attacks may have devastating consequences.

Fanaticism inspired by all kinds of religious-sectarian-nationalist convictions is taking on a millenarian tone, with small sectarian groups that lack clear political or social agendas other than destroying the civilisation. While the traditional terrorist movements historically consisted of hundreds of members, the new terrorist groups can be very small, consisting of a few people. The smaller the group the more radical it is likely to be, the more divorced from rational thought and the more difficult to detect and impossible to infiltrate. Some believe it is unlikely that the terrorist possess technological know-how and the resources to make use of WMD. But the technological skills required are not that complex and resources needed not that rare or expensive. It is possible that government of states of concern, which may themselves not use these weapons for fear of retaliation can readily supply the raw material or the finished product to terrorist or the terrorist-supporting states either by political design or for monetary gains.

Some believe that the horrific consequences of using WMD will deter even the most hardcore from using them. But this underrates the element of blind aggression, of rage, of suicidal impulses, of sheer madness, which unfortunately has always been part of human nature. The ready availability of WMD has now come to pass. The character of terrorism is changing, any restraints that existed are disappearing and above all, the threat to human life has become infinitely greater. Traditional terrorist attacks have generally included high profile bombing attacks, but now they appear to be growing more destructive in nature. Religiously motivated terrorism is becoming ubiquitous, followed by politically motivated and even millennial-based terrorism. The Cold War rivalry that held satellite states and ethnic groups together has changed. To make matters worse the Soviet collapse has destabilised the security of an estimated thirty thousand nuclear weapons, hundreds of tons of fissile material and radioactive waste, more than forty thousand metric tons of chemical and biological weapons, huge stockpiles of sophisticated conventional weapons and thousands of scientists with the knowledge to produce WMD. Proliferates such as North Korea, Pakistan and the People's Republic of China (PRC) exchange WMD technology and production facilities with pariah states known to sponsor terrorism.

More than twenty-five countries have, or may be developing WMD and the means to deliver them, a larger number are capable of producing such weapons. In addition, the NBC proliferation threat has become transnational and now has the potential to emerge from terrorist organisations or organised crime groups. WMD share common connections to one another that may be attractive for terrorist organisations. The first is the immense lethality. A single weapon can kill thousands of people or be used on one or several individuals. The second is portability, which allows them to be easily delivered against civilian populations and inadequately prepared military forces. The third is accessibility, which means that they may fall into terrorist hands, despite the best efforts at prevention. These three characteristics combine to make WMD one of the single most serious long-term security threats. Thus, WMD terrorism is an international problem, one without borders that demands a solution on an international scale and terrorists are known to be informed and capable. The techniques and materials associated with acquisition programs of WMD are becoming more accessible and within reach for well-connected terrorist organisations.

In the prevailing security situation, due to manifestation of terrorist groups spread over a wide area, India is as vulnerable as other developed countries to an attack by terrorists using weapons of mass destruction. Certain countries in India's neighborhood propagate state sponsored terrorism as an instrument of their foreign policy. The region is beseeched with large number of problems related to poverty, unemployment, underdevelopment, and population explosion. These underlying conditions pave way to the manifestation of insurgencies/terrorism and border disputes among the states. Given the international pressure on all nations prohibiting chemical and biological warfare agents, threat of overt use of such WMD's has given way to their covert use by non-state parties. Due to spread of dual-use technology and expertise, the prospects of chemical, biological and radiological terrorism have increased. The relative ease of production and the cost factor has made these weapons more attractive to the terrorists.

The question of potential use by terrorists is an important and timely one, especially when considered in the light of the explosion in information availability and growing concerns over proliferation and material control issues. The possible level of physical destruction, injury and loss of life from these weapons is so great in and of itself that the potential for terrorist acquisition and use warrants serious consideration. The expected consequences demand serious consideration of the potential for WMD use by terrorists. Identifying and accurately addressing the threat potential have become important components of planning, prevention and consequence management efforts.

Developed countries have adopted a comprehensive approach to counter the growing threat from weapons of mass destruction. This strategy is integral to any security strategy. In India's case, there is a need to also develop strategy for consequence management of WMD terrorism. There is a need to shift our strategic focus to the threats. We as a nation today face and lay more emphasis on dealing with insurgencies, terrorist networks and other non-traditional threats. We need to reorganise ourselves to deal with the threat of terrorists using weapons of mass destruction. To succeed, we must take full advantage of today's opportunities including application of new technologies, increased emphasis on intelligence collection, collation and analysis and strengthening of new partnerships a..d relationship with other countries.

# 2

## Terrorism

### A Security Challenge

*A nation is secure to the extent to which it is not in danger of having to sacrifice core values, if it wishes to avoid war, and is able, if challenged, to maintain them by victory in such a war.*

**Walter Lippmann**

Until the end of the Cold War, a bipolar world maintained international security through a combined system of alliances, spheres of influence, and global and regional multilateral institutions. With the disintegration of the Soviet empire, however, a uni-polar world has emerged that has dramatically changed the landscape of international security. The end of the Cold War has also permitted the emergence into free, independent societies of many formerly suppressed countries and people. It has also awakened old ethnic conflicts and cultural disputes that, as a result of the restraints imposed by superpower rivalries, had lain dormant, both between and within nations and societies. The 11 September 2001 attacks on Twin towers in America, the only Superpower, has added a new dimension. Perpetrators of these attacks belonged to different countries. They planned and executed the attacks with precision, killing thousands of people. Such terrorist's suicide teams are prepared to sacrifice their lives using the most unconventional weapons.

Many authors have tried to define terrorism in their own way. The essence of all definitions is that targets of terrorism are not the victims but the governments, the public or constituents among whom the terrorists want to engender a reaction such as fear, repulsion, intimidation, overreaction or radicalisation. Although it is very difficult to define terrorism, certain common aspects are central to the concept of terrorism[1]. First, it is political in nature aimed at bringing political change. Second, no matter how well organised the terrorist groups are, they are always non-state actors. Terrorism also thrives on political and economic support from different sources like state forces. Inspite of this, their non-state characteristics set them apart. Third, terrorism deliberately targets innocent civilians. Fourth, use of force by state is subject to international laws or norms and conventions; terrorists do not abide by international norms. At the same time, when states use force against innocent civilians like the case of the massacre of Tutsis and moderate Hutus in April 1994, in the aftermath of death of the Rwandan President, it has to be treated as an act of terrorism. Therefore, multiple terrorists' acts perpetrated by regimes against their own populations will have to exclude from this proposition of authorised or legitimate use of force by state. Examples of terrorist acts such as the kidnapping of foreign tourists in Kashmir by militants and in Philippines by Abu Sayyaf group in April 2000, bombing of the US Embassy in Kenya and Tanzania and the 11 September 2001 attacks on America illustrate these special attributes of terrorism. According to Andrey Kurth Cromin, terrorism is the threat or use of seemingly random violence against innocents for political ends by non-state actors. The problem with terrorism is its definition, one that is acceptable to all states and parties in a dispute. Most states do not consider terrorism as an evil or as socially harmful. They are instead regarded as freedom fighters.

Several definitions of terrorism are available in the literature. Some focus on the perpetrators, others on their purposes and still others on ´their techniques. Two critical characteristics are, firstly, terrorism is aimed at non-combatants and, secondly, terrorists use violence to instill chronic fear in the targeted population. Therefore, terrorism is an activity that employs or threatens to employ premeditated violence to instill fear in a victim in pursuit of strategic goals specified by the perpetrator.

Jessica Stern defines terrorism as an act or threat of violence against non-combatants with the objective of exacting revenge, intimidating, or otherwise influencing an audience. This definition avoids limiting perpetrators or their purpose.[2] It allows for a range of possible actors (states or their surrogates, international groups, or a single individual) for all putative goals (political, religious or economic) and for murder for its own sake.[3] Asma Jahangir provides another view of terrorism. The noted Pakistani human rights activist has defined terrorism as organised violence by small groups against the state for political purposes. The aspects of it being organised, requiring conscious planning and direction, undertaken by small, usually secret bodies of armed men and women, its object being the state though the victim could be the common citizens, and the purpose political.[4]

The Government of India vide Section 3(1) of the Terrorist and Disruptive Activities (TADA) Prevention Act, 1985, defines a terrorist act as "Whoever with intent to overawe the government as by law established, or to strike terror in the people or any section of people or alienate any section of people or to adversely affect the harmony amongst different sections of the people does any act or thing by using bombs, dynamite or other explosive substances or inflammable substances or firearms or other weapons or poison or noxious gases or other chemicals or other substance (whether biological or otherwise) of a hazardous nature in such a manner as to cause, or is likely to cause, death of, or injuries to any person or persons or damage to, or destruction of property or disruption of supplies or services essential to the life of the community, commits a terrorist act". *Encyclopedia Britannica* defines terrorism as "the systematic use of terror - such as bombing, killing and kidnapping - as a means of forcing some political, social or religious objectives by relatively small group, devoid of popular support".[5]

Paul Wilkinson identified five distinguishing characteristics of terrorism:[6]

(a) It is designed to create a climate of extreme fear.

(b) It is directed at a wider target than the immediate victims.

(c) It inherently involves attacks on random or symbolic targets, including civilians.

(d) It is considered by society in which it occurs as 'extra-normal', as it violates the norms regulating disputes, protests and dissent.

(e) It is used primarily, though not exclusively, to influence the political behavior of factions governments, communities or specific social groups.

Terrorism as defined by the Federal Bureau of Investigation of US states: "The calculated use of unlawful violence or threat of violence to inculcate fear, intended to coerce or to intimidate governments or societies in the pursuit of goals that are generally political, religious or ideological".[7] The types of force employed by terrorists can be diverse and include devices within the major classification of weapon of mass destruction. The classifications are chemical, biological, radiological, nuclear and high yield conventional explosives. WMD terrorism involves the most modern and the most extreme form of random violence. Nuclear, chemical and biological weapons are inherently terrifying; in most scenarios for their use, the fear they would cause would dwarf the injury and death. Dread of these weapons creates its own dangers, if victims panic and attempt to flee, they may spread contamination and disease, thus furthering the area of impact and increasing the success of the terrorists.

## History of Terrorism

The chequered history of traditional terrorism may be said to have begun in the First Century BC, till it transferred to modern terrorism, during the French Revolution. The word terrorism actually comes from the Latin word meaning fear but the word terrorism first came into widespread use during the French Revolution. During the period of the revolution, over 300,000 people were arbitrarily arrested and 17,000 were executed without trial. The first definition of terrorism in the *Oxford English Dictionary* refers to government by intimidation as carried out by the party in power.[8]

Terrorism has evolved over the centuries. Its early acts are of assassination, regicide and tyrannicide. Three historical terrorist organisations are of particular interest here, the Zealots-Sicarri, the Assassins and the Thugs. All were inspired by religious conviction, and all were active internationally. Their weapons were primitive – the sword, the dagger and the noose – and yet they were more destructive than any modern group.[9] The Sicarri, an extreme Jewish faction, were active after the Roman occupation of Palestine in the first century. The group began by murdering individual victims with short (sica) daggers and swords hidden under their coats and later they turned to open warfare with the objective of creating a mass uprising against the Greeks in Judea and against the Romans, who governed both Greeks and Jews. There seems to have been a social element as well; their attacks were invariably directed against moneylenders. The revolt had unforeseen and devastating consequences, leading to the killing of high priests and destruction of Jewish temples and to the mass suicide at Masada. Later revolts inspired by the Zealots-Sicarri led to the extermination of the Jews in Egypt and Cyprus, the virtual depopulation of Judea and the Exile. This was to remain the central feature of the Jewish experience for the next two thousand years.

Another early example of a terrorist is the Order of the Assassins in the eleventh century, an offshoot of the Ismailis, a Muslim sect that operated from 1090 to 1275. Hassan 1 Sabah, the founder of the order, was born in Qom, the Shiite centre in northern Persia. Sabah adopted an extreme form of Ismaili doctrine that called for the seizure of mountain fortresses. Years later the Assassins decided to transfer their activities from remote mountain regions to the main urban centres. Their first urban victim was the Chief Minister of the Sultan of Baghdad, a Sunnite by religious persuasion and, therefore, an enemy. During the years that followed, Assassins were active in Persia, Syria and Palestine, killing a large number of enemies, mainly Sunnis but also Christians including the ruler of the kingdom of Jerusalem. There was a great deal of mystery about this movement and the dissimulation used. Seen in retrospect, the impact of the Assassins was small since they did not make any converts outside their mountain fortresses, nor did they produce any significant changes in Muslim thought or practice. However, their main contribution was perhaps the art of attacking, disguised – taqfir, or deception – as a devout emissary but in fact on a suicide mission in exchange for which they were guaranteed the joys of paradise.

The Thugs were active in India for at least six hundred years, until the late 19th century. No one knows precisely how many people they killed, but estimates range over 500,000. Their victims were travellers, whom they strangled, dismembered, buried and robbed. The Thugs' principal objective was to provide sacrificial offerings to the goddess Kali. They strove to prolong the death agony of the victims, believing Kali enjoyed their victim's terror.

The 19th century, also witnessed the emergence of guerrilla warfare and terrorism. Guerrilla warfare appeared first in the framework of the Napoleonic wars in Spain and Russia, and continued in various parts of Asia and Africa, and reached its high tide after the Second World War, with the disintegration of the European Empires. Terrorism as we know it grew out of the secret societies of Italian and Irish patriots but it also manifested itself in most Balkan countries in Turkey and Egypt, and of course among the extreme Anarchists, who believed in the strategy of propaganda by deed.[10] In Russia, at the turn of the century, the intellectual elite considered terrorist violence as the only effective way to modernise Russian society. Prior to the First World War, the Russian terrorists were by far the most active and successful. Terrorism was widely discussed among the European Far Left, not because the use of violence as political statement was a monopoly of the left but because the Right was the political establishment, and prior to World War 1 the Left was the agent of change, trying to overthrow the party in power. However most leaders of the Left rejected terrorism for both philosophical and practiced reasons. They favoured collective action, such as strikes, demonstrations, perhaps even insurgency but neither Marx nor the anti-Marxists of the left believed in the "philosophy of the bomb"[11]. They gave political support to the Irish Patriots and the Russian revolutionaries without necessarily embracing their tactics.

The late 20th century also saw a resurgence of holy terror – the kind practiced by the Zealots-Sicarri, the Assassins and the Thugs. Religiously inspired groups tend to be more violent than their secular counterparts. Religiously motivated terrorists might decide to use WMD, particularly biological agents, in the belief that they may be emulating God. The fifth plague with which God punishes the Pharaoh in the story of the Israelites' Exodus from Egypt is murrain, a group of cattle diseases that includes anthrax (Exodus 9.1-6). In I Samuel 5:9, God turns against the Philistines with a very great destruction, killing them with bubonic plague. Some terrorists might feel they were following God's example by employing these agents. The prime example was the massacre of eleven Israeli athletes in 1972 Munich Olympics, a first of its kind followed by a series of incidents like airlines hijacking and hostage taking. A new kind of terrorism emerged in the second half of the twentieth century: random violence. Especially since the 1970s, terrorists have increasingly targeted people who had no obvious connection to the terrorists' grievances.

Another possible motivation for terrorists' using WMD is pre-millennial tension. The millenarian idea is that the present age is corrupt and that a new age – the millennium – will dawn after a cleansing apocalypse. Only a lucky few will survive the apocalypse and experience paradise. Most millenarians believe that God is responsible for determining the date of the apocalypse, which is a pre-condition for the appearance of the Messiah. But some believe that human can speed up the process along. Prayer, repentance and martyrdom are common techniques; but some millenarians may add terrorism to this list.

Terrorists might also scale out WMD to commit economic terrorism. Unlike conventional weapons, radiological, chemical and biological agents could be used to destroy crops, poison foods or contaminate pharmaceuticals. They could also be used to kill livestock. Such acts could inflict enormous economic and political costs on the target country's industry or company.

## State Terrorism

State sponsored terrorism, warfare by proxy, is as old as the history of military conflict. It was an established practice in ancient times in the oriental empires, in Rome, in Asia and

Europe. No empire, however powerful could afford to live in a state of perpetual war with its neighbours. There was a cheaper and less risky alternative i.e. to support dissenters, separatists, ambitious politicians or simply malcontents inside a rival state. This strategy was worked out as a defensive means by a weak state to forestall the aggressive designs of a powerful potential enemy. At times this was the strategy adopted by the powerful states against its weak opponent as part of its offensive plans to weaken the state and also to prepare grounds for an invasion. There are innumerable examples of state sponsored terrorism in modern days. The Western powers have supported all kinds of nationalist and anti-Communist rebels, without much success against the Soviet Union. A similar activity is being promoted by Pakistan against India.

In the 1970s and 1980s Libya was one of the foremost sponsors of international terrorism. In 1969, in a coup, Colonel Qadhafi came to power. Qadhafi's ambition was to spearhead an Arab-Islamic revolution in which he saw himself not only as the chief ideologist (by virtue of his "Green Book") but also as chief strategist.[12] Libya was able to support financially a variety of terrorist activities due to its income from oil. Libya provided support not only to Arab terrorist groups, variety of Central and West Asian groups but also to terrorists from Ireland and Philippines. It trained foreign terrorists on its soil in various terrorist camps. It also was able to construct various factories to produce poison gas and other chemical weapons. (Osama Bin Laden has been one of the initial beneficiaries of his largesse). Over a period of time, Libya under Qadhafi has renounced terrorism and has been making efforts to cosy up to the Americans.

Another Arab country that has played an active role in sponsoring terrorism has been Sudan. The Sudanese government collaborated closely with Libya and has been assisting various terrorists groups all over the world. It granted all the facilities of its training camps to Osama bin Laden; a billionaire Saudi businessman who has been responsible for many terrorist attacks in his native country and others in Middle East, and subsequently who shifted base to Afghanistan and has been the chief architect of the 11 September attacks on America.

Even democratic states have on occasions engaged in terrorists operations, or at least contemplated such actions. The United States at one stage considered terrorist actions against Cuba and Castro in person, but these plans, usually half-baked, were seldom attempted. The Israelis systematically hunted down the perpetrators of the massacre in the Munich Olympics village in 1972 and they have sometimes attacked leading members of Palestinian terrorist organisations.

Seen in long-term perspective, state-sponsored terrorism has been on the decline, in as much as the numbers of action have declined. But it has by no means vanished. Those who have dropped out are countries vulnerable to international pressure. In the post cold war era, which is plagued by mostly intra-state conflicts involving multiethnic groups, managements of such complex situation requires multifunctional approach.

Terrorism is a transforming phenomenon. Its escalatory nature is evident with groups acquiring greater capability for destruction with each year of existence. Bruce Hoffman has carried out an analysis of the difference in old and new forms of terrorism, based on factors such as organisation, terrorist acts, lethality, motivation and state sponsorship.[13]

**Table 2.1 Analysis of Old and New Terrorism**

| | **Old Terrorism** | **New Terrorism** |
|---|---|---|
| **Organisation** | Identifiable organisation, clear command and control and a set of political, social and/or economic objectives. | Lack of specific existing or identifiable organisation. Networked to a myriad number of ethnic ideological and criminal groups. |
| **Acts** | Selective and discriminate acts of violence | Indiscriminate acts of violence. |
| **Size of Organisation** | Small | Large. (The Al Queda can call up to 40,000 -50,000 well-trained fighters in the Muslim world.) |

(Contd.)

| | Old Terrorism | New Terrorism |
|---|---|---|
| **Lethality** | Limited | Increased.(In the 11 September 2001 attacks, almost 3,000 people were killed). |
| **Motivation** | Ideological | Religious-criminal |
| **State sponsorship** | Limited or negligible | High level of state sponsorship. Terrorists can be called "pseudo state forces". |

The changing nature of terrorism has also been reflected in the analysis of the summary of casualties due to terrorist violence in the 1970s, 1980s and between 1990-1996. There has been a sharp increase in deaths and injuries in terrorist incidents within a short span of two decades, which is alarming and demonstrates the lethality of modern terror[14].

**Table 2.2. Analysis of Casualties due to Terrorist Violence**

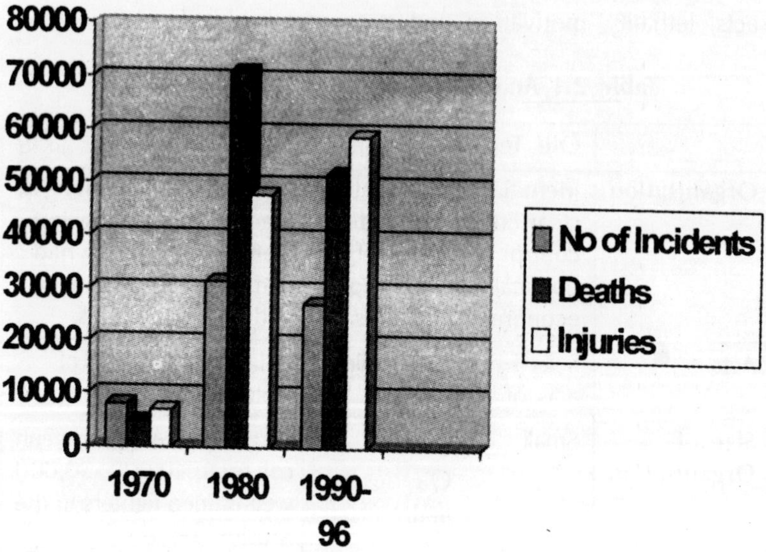

The reasons for the growth of modern terrorism as per Ruby Maloni, are as follows[15]:

(a) Nature of modern civilisation where a technological society has led to the growth of targets such as transportation centres and means of public transportation to include air, which are not only highly vulnerable but are seen as a symbol of the elite. The dispersion of targets in a modern metropolis renders them highly vulnerable, providing adequate pportunity for strikes to the dissenter.

(b) Easy availability of weapons, including those of mass destruction.

(c) Rise of the sub-national group capable of modern combat.

(d) Global and transnational networking due to extensive communication and transportation facilities available worldwide.

(e) Media proliferation providing the terrorist a medium for propagating the philosophy of his acts.

(f) Growth of liberalism, which proscribes coercive measures against terrorists who are categorized as misguided youths affected by social, economic and political vagaries. The influence of human rights groups such as Amnesty International and Human Rights Watch, while ensuring greater liberty, has also restricted tough actions by states against terrorists.

## The Terrorist Groups

The growth of transnationalism is most apparent in the field of terrorism with terrorist groups networking to emerge as powerful players in their own sphere of influence. The deadliest terrorist groups in the world today are those who have the widest and most secure networks, be it socio-political, finances, drugs or arms. The al Qaeda is an umbrella organisation established by Saudi millionaire Osama bin Laden to unite Sunni Islamic extremists. Its goal is to establish a pan-Islamic Caliphate throughout the world, by overthrowing regimes it deems "non-

Islamic". It issued a statement in February 1998 saying it was the duty of all Muslims to kill US citizens, civilians or military, everywhere. The most powerful and deadly of all terrorist groups, with an international reach, it has linkages with other terrorist groups in all parts of the world.

The *Liberation Tigers of Tamil Ealam* (LTTE) began their armed conflict with the Sri Lankan government in 1983. The exact strength of the cadres of the LTTE has never been revealed but is estimated to be upto 10,000 fighters in Sri Lanka. It is involved in narcotics smuggling and has a worldwide presence. Known for its ruthless suicide bomb attacks, the LTTE has been in partial control of parts of Sri Lanka from time to time. Recently, it has offered ceasefire, mainly to deflect western pressure on terrorism after the 11 September attacks.

Hamas is the ideological mooring for all extremists of the Palestinian group. However, it is said to have been undermined by internal conflict and diminished external support. It does not recognize Israel's right to exist and has mainly carried out attacks against the state. Its active cadre strength is unknown but it has ten of thousands of supporters across the world.

Foreign terrorist groups such as the Lashkar-e-Taiyaba and Jaish-e-Mohammad, operating in Kashmir, including their aliases, are increasingly acquiring greater potential for destruction and lethality. Though not in the same league as the LTTE so far, their brutality and indiscriminate use of violence make them especially fearsome.

### *Terrorism in South Asia*

Asia is likely to remain the main arena of international terrorist operations in the foreseeable future. South Asia and contiguous regions has been target of terrorism, insurgencies, and separatist movements much longer compared with other parts of the world. Afghanistan, Pakistan, Central Asia, India, Sri Lanka and Nepal are facing serious terrorist threats. Internecine ethnic, communal and separatist conflicts pose a serious threat to stability and interstate relations in this region.

South Asian region extends from Pakistan in the West to Bangladesh in the East, Nepal to the North and Sri Lanka to the South, with India forming the geographical as well as the identity core of the region. These countries of the South Asian region had achieved independence over six decades ago, a period that has been most turbulent for the people of the region. The peripheral states of the South Asian region have also been in turmoil, be it Afghanistan in the West, Central Asian Republics and Tibet in the North and Myanmar in the East. It represents one of the most conflict-ridden areas in the world with a number of interstate problems. A number of issues like disagreements on mutually acceptable frontiers, water sharing, growth of trans-border migration, cross border terrorism, rise of fundamentalism and the curse of underdevelopment bedevil the South Asia conflict cauldron.[16] Ethno-nationalism and border disputes have been dominant issues in South Asian Security in the past few decades.

South Asia and contiguous regions has been the target of terrorism, insurgencies and separatists' movements due to underlying conditions such as poverty, corruption, over population, religious conflicts, and ethnic strife. Terrorists have been effectively able to exploit these conditions, thus endangering the very survival of the nation states. Continuous violence in the region has also demonstrated a wider range of effective instrumentalities of sub-conventional warfare.

The power of terror and of such pattern of sub-conventional warfare is evidenced in a wide range of conflicts in South Asia and it is useful to take an overview of the sheer expanse of disorder and violence with which the region is confronted.

### *Ideological Streams*

Within South Asia, three broad streams of justifications underlie all terrorists and insurgent violence. The most destructive of these are in evidence in the two high intensity conflicts in the region - in Jammu and Kashmir and Nepal and these are the ideologies of political Islamist Jihad and of Maoist Left Wing extremism. But these two theatres are only the most visible manifestations of a much wider phenomenon. Both are traditional ideologies that refuse to recognize national boundaries and international conventions. The third stream is of a wide range of ethnicity-based insurgencies, particularly in India's Northeast and the Liberation Tigers of Tamil Eelam (LTTE) movement in Sri Lanka. Despite wide ideological divergences in these three streams, there are significant conceptual and operational overlaps.

Islamism is currently the most powerful and widespread ideological stream sustaining terrorism in South Asia, and strong linkages exist with violent Islamist ideologies and movements across the world. Within the region, the hub of Islamist terrorism has long been the militarised Islamist fundamentalist state of Pakistan. As a region, South Asia has a long history, both of communal confrontation and violence, on one hand, and of co-existence within an electic culture that has accepted differences on the other. This said, it must also be emphasized in the strongest terms possible that moderate Islam is today, under deep, penetrating and sustained attack in every concentration of Muslim population throughout South Asia, and there is 'Hardening' of beliefs that may lend itself to the extremist jihad in an uncertain future.

The 'footprints' of a majority of key acts of international Islamist terrorism invariably pass through Pakistan, right from 11 September where virtually all the participants had trained, resided or met in, coordinated with or received funding from or through Pakistan – to major acts of terrorism across South Asia and South East Asia, as well as major networks of terror that have been discovered in Europe. Pakistan has harvested an enormous price for its supposed cooperation with the US, and in that it has combined deception and blackmail including nuclear blackmail to secure a continuous stream of concessions. Pakistan's case for incremental aid has been that if it does not receive extraordinary dispensations and indulgences, it will in effect 'implode' and in the process do extraordinary harm to others. Part of the threat of this implosion is also the spectre of the transfer of its nuclear arsenal and capabilities to more intransigent and irrational elements of the Islamist far right in Pakistan whose interests in terrorism are strategic.

Pakistan has consistently projected Kashmir as the 'core issue' of conflict in South Asia. The fact, however is that the conflict in South Asia cannot be resolved within Kashmir, since the confrontation between the two nations is not about territory, but it is an irreducible ideological confrontation between a pluralist secular democracy and a religious fundamentalist, intolerant and exclusionary ideology.

Within Pakistan a number of internal contradictions are creating increasing stresses. The efforts to orchestrate a transition to democracy through a controlled military regime has immensely weakened democratic and secular forces in Pakistan, even as it has further entrenched the military-jihadi-feudal combine of revanchist forces in the country.

Bangladeshi soil is increasingly being used for terrorist and for subversive activities by religious extremists, pan-Islamist terrorist groups and insurgents operating in India's Northeast, substantially through the active collusion of Pakistan's Inter Services Intelligence, (ISI) and Bangladesh's Directorate General of Field Intelligence (DGFI). ISI has been able to make major inroads into Bangladesh by harnessing past linkages with Islamist fundamentalists and certain sections of the military and political establishment. Internally also, various Islamist groups remain active and have enlarged their subversive agenda in alliance with a number of transnational Islamist terrorist groups, validating perceptions that the country is emerging as a major safe haven for Islamist terrorists formations.

Various terrorists groups operating in India's Northeast continue to find safe haven and operating bases in Bangladesh territory. Groups like the United Liberation Front of Assam (ULFA) have found regular support from official channels for carrying out anti-India activities. Leaders of ULFA have established direct business linkages including partnerships in corporations and financial operations that are run by, or co-owned with leaders of such terrorists' organisations.

Jammu and Kashmir (J&K) is currently the most significant internal security challenge faced by the country, and four Islamist fundamentalist groups, all of them headquartered in Pakistan, constitute the gravest threat in the State: Lashkar-e-Taiyaba (LeT) now renamed Jamaat-ud-Dawa; Hizb-ul-Mujahideen (HM); Harkat-ul-Mujahideen and Jaish-e-Mohammad (JeM) now renamed Khuddamul Islam. All these groups seek integration of J&K with Pakistan. The JeM and the LeT are both closely connected with the al Qaeda and are members of Bin Laden's International Islamic Front (IIF) for Jehad against Jews and Crusaders. The HM is the largest group in the State, accounting for up to 60 percent of the total terrorists cadres in the state.

The Islamist groups operating with Pakistan support in Kashmir do have a clear pan-Islamist agenda and are known to have created a network of terrorists cells in a number of Indian states with the intention of provoking communal confrontation, engineer terrorists incidents and recruit soldiers for pan-Islamic Jihad. The networks and support structures of these organisations were painstakingly constructed by al Qaeda- ISI combine, backed by enormous flow of financial support from West Asia in a sustained strategy of 'erosion, encirclement and penetration'.

Nepal, an infant democracy is also being challenged by rising extremist Islamist activity, despite the fact that Muslims constitute a bare three percent of the population. Such activity is presently directed against India with the active support of ISI involving strong organised criminal networks and prominent political leaders. The Indian region of Terai adjoining Nepal border has seen increased activity of construction of mosques and madrassas with financial flow from Pakistan, Saudi Arabia and from a range of pan-Islamic organisations. Simultaneously there has also been proliferation of fundamentalist organisations, with the result Nepal is, consequently emerging as an important staging post for Pakistan's strategy of penetration, erosion and encirclement, and is increasingly the preferred route for terrorists' movements to various areas of low-intensity conflict in J&K and the Northeast.

### *Left Wing Extremism*

There are increasing trends towards escalation and widening of the sphere of Left-Wing extremist violence in Nepal and India as well as of deepening linkages between Left-Wing insurgents in the two countries and beyond. Cadres of the

Communist Party of Nepal (Maoist) (CPN-M) are increasingly working closely with the Naxalite groups in India particularly the Marxist Communist Centre (MCC) and the Peoples War Group (PWG). These Naxalite Groups have long been active in parts of the States of Andhra Pradesh, Jharkhand, Bihar, Chhattisgarh, West Bengal, Orissa, Madhya Pradesh, Uttar Pradesh and Maharashtra. Apart from their traditional strongholds, there has been significant expansion of Naxalite activities into new areas and they now have presence and varying degrees of activity in 74 Districts across 12 Indian states.[17]

The Maoists in Nepal have overrun virtually all the districts of the country, though they have suffered apparently grievous losses since the collapse of peace talks with the government in August 2003. A close analysis of developments reveals the spectacle, rather of a country hurtling towards chaos, with the state and its agencies in headlong flight. With the exception of the Army and paramilitary forces, Kathmandu has no presence in increasing parts of the country. More significantly, there is little permanent presence of enforcement agencies in the rural areas, as the police abandon the smaller stations to consolidate at the district headquarters. The resultant disorder is the insurgent's ally, and the enemy of the state. In the absence of the restoration, strengthening and extension of the permanent institutions of governance, including the critical institution for the maintenance of law and order - the police station - no permanent resolution to the problem of insurgency in Nepal seems possible.

There are indicators that the successes of the Maoists in Nepal could become a model for left wing extremist movements in other parts of South Asia.

### *Ethnicity*

Ethnicity-based insurgencies are endemic in India's Northeast, with Assam, Manipur and Tripura as the worst affected states. A ceasefire exists between the Government and the two principal insurgent groups in Nagaland, and a negotiated solution is being sought to the insurgency through peace talks. The states of Meghalaya, Arunachal Pradesh and Mizoram are affected with lesser insurgencies relating to the problems of 'overflow' of movements from neighbouring states. In the whole region of Northeast there is a vast proliferation of ethnic insurgent groupings representing progressively narrower tribal interests. The insurgencies in the region are compounded by, and often sourced in the demographic destabilisation that has continued virtually unchecked over the past century, and which has assumed particularly ominous overtones with the redrawing of political maps after India's partition.

The border management problem in the Northeast is gigantic, especially the Bangladesh border of 4,095 Kilometres being the most intractable problem. Illegal migration, the existence of an elaborate terrorists infrastructure in safe havens across the borders, the growth and entrenchment of organised criminal gangs and syndicates with powerful political and communal influence and patronage along the border, and a strengthening network of well funded institutions for the communal mobilisation of the migrant community particularly through a growing complex of mosques and madrassas – are some of the most dangerous trends that are required to be contended with in the region.

Sri Lanka is faced with the problem of bringing about a peaceful settlement of the 'Tamil Question'. A ceasefire between the Government and the Liberation Tigers of Tamil Eelam (LTTE) is holding on, but there has been no progress on the peace talks. Even as no available framework exists and concomitantly the squabbles among the Sinhala political parties continue, the LTTE is emerging as a far superior force as compared to the Sri Lankan Army. The organisation's infrastructure, military as well as financial, still exists, and has in fact, been consolidated further during the peace process period. The presence of a majority Muslim population, deprived of a share in the decision making process, in the Eastern province, where the LTTE is now ruling, constitutes a potential tinder box for Islamist extremism.

The Sri Lankan situation also has potential negative fallout in the Indian state of Tamil Nadu, in case the current peace process results in a quasi-independent administrative or state structure in the Northern and Eastern provinces dominated by the LTTE.

Despite the apparently irreconcilable differences between the three ideological streams that underline the terrorists and insurgent violence in South Asia, they share a number of characteristics: they are authoritarians in outlook and reject liberal democracy and the political economy of the globalisation process; they are movements largely located among population - not necessarily poor- that have been ignored or marginalised by the process of modernity and that, in fact, seek smaller worlds within borders that will seal them off from modernity. These ideologies are consequently, in irreducible conflict with the liberal democratic traditions.

However within South Asia, economic trends are beginning to gain centrality in terms of projected futures, and as Indian economy strengthens, military and economic disincentives will increasingly come into existence for the sponsorship of terrorism by Pakistan and Bangladesh, with a dampening effect on long-term trends in terrorism. The benefits of interstate economic investments and trade are, however, unlikely to be transferable to the security domain in the near term. The processes of globalisation and the structural adjustments within the regional economies, moreover, may exacerbate social and political tensions, especially within the segments of the population who are marginalised by these processes.

## References

1. Andrey Kurth Cromin, "Behind the Curve; Globalisation and International Terrorism", *International Security,* Vol 2, No. 3 (Winter 2002/03), p.49.
2. Jessica Stern, *The Ultimate Terrorists* (London: Harvard University Press, 1999), p.11.
3. Martha Crenshaw, *Terrorism and International Cooperation* (New York: Institute for East West Security Studies, 1989), p.5.
4. Asma Jahangir, *States of Violence: Nature of Terrorism and Guerrilla Warfare* (New Delhi: Dominant Publishers and Distributors, 2000), p.79.

5. *Encyclopedia Britannica*, London, Volume VIII, pp.310-311.
6. Grant Wardlaw. *Political Terrorism: Theory, Tactics and Countermeasures.* (Cambridge University Press, 1989), p.76.
7. Jeffrey A. Adams and Stephen Marquette, F*irst Responders Guide to WMD* (Alexandria: American Society for Industrial Security), p.120.
8. Stern, n.2, p.13.
9. David C Rapoport, *Fear and Trembling: Terrorism in Three Religions Traditions, American Political Science Review 78,* No.3 (September 1984).
10. Walter Laqueur, *The New Terrorism* (London: Phoenix Press, 2001), p. 15.
11. Stern, n.2, pp. 70-71.
12. Laqueur, n.10, p.22.
13. Bruce Hoffman, *Terrorism Trends and Prospects* (Santa Monica, CA: Rand, 1999), pp. 8-10.
14. Stern, n.2. p.6
15. Ruby Maloni, *Historical Perspectives and Future Implications of International Terrorism* (New Delhi: Kanishka, 2001), pp.30-32.
16. Harshe Rajan "*Understanding conflicts in South Asia: Inter and Intra State Conflicts*" in Sudhir Jacob George (ed), South Asia (New Delhi: South Asian Publishers, 2001), p. 19.
17. Aditya Sinha, "The Mess in Our Neighbourhood" *Times of India,* 17 0ctober, 2004. p.16.

# 3

# WMD Technologies and Threats

*In our life times, the morally savage, unruly killer, not the trained, disciplined soldier, will be type of enemy most frequently encountered by Euro-American militarism.*

**Peters**

Weapons of mass destruction – nuclear, biological and chemical – have long been a subject of human imagination, in both the literary and literal sense. These agents are inherently terrorising and evoke moral dread. Biological and chemical weapons have been used in wars for centuries and nuclear weapons that have been used only twice lead large-scale casualties and destruction. The state attacked with such weapons would have difficulty in controlling panic and fear. A weapon of mass destruction is classified as one which has the capability to produce a great deal of devastation, which can produce destruction of a far greater magnitude than the level of effort that goes into its production and delivery, which has indiscriminate effect, which is widespread, generalised and hard to defend.[1] These weapons are capable of killing many people at the same time. The use of WMD terror is all the more threatening in the modern world, as crowded targets in urban areas are vulnerable to WMD attacks. Thus, while "normal" or non-WMD terrorism relying on conventional explosives would cause casualties in the low hundreds, terrorism, relying on chemical and biological or even nuclear weapons, will cause casualties in the low thousands and upwards.[2] What is interesting to note is the reason why the nuclear, biological and chemical weapons have been grouped together. The answer lies in their common characteristics: -

(a) All produce mass casualties.

(b) Weather and terrain attenuate the effects of weapons.

(c) International abhorrence to the use of the nuclear, biological and chemical weapons.

(d) Due to their impact, the political leadership tries to control the employment, deployment and storage of these weapons.

Title 10, US Code, Section 1403 defines WMD as "Weapons or devices intended, or having the capability, to cause death or serious bodily injury to a number of people through the release, dissemination, or impact of toxic poisonous chemicals or their precursors; a disease organism; or radiation."[3] The general characteristics of WMD are shown in the Table below:

**Table 3.1 General Characteristics of Weapons of Mass Destruction**

| Characteristics | Nuclear | Chemical | Biological |
|---|---|---|---|
| Area Affected | Large | Relatively small | Potentially very large |
| Detectability | Unmistakable | Difficult | Very Difficult |
| Time to Detect and Identify | Near Real Time | Seconds | Minutes to Hours |
| Time from Exposure Until Onset of Effects | Varies with Dosage | Normally Minutes | Normally Days |
| Medical Treatment | Very Limited | Limited | Effective if Timely |

Source: JCS J-3 *Weapons of Mass Destruction (WMD) Handbook*, Washington, DC (U.S. Department of Defense, February 2001), p.3.

Chemical and biological agents commonly known as the poor man's nuclear weapons are becoming an increasing threat. The use of a military agent in Japan signified the introduction of weapons of mass destruction into the terrorists' arsenals. An alarming resource for many types of bombs is the Internet, which gives small groups a wealth of information on preparing many types of weapons of mass destruction. What is even more alarming is the demonstrated willingness to use them. Although no major act of use of nuclear, chemical, and biological agents by terrorist groups has yet surfaced, the prospect of an attack on a nuclear power plant (causing the release of nuclear material) and biological terrorism is chilling. NBC incidents are closely related in terms of impact. All NBC attacks have a powerful psychological effect and result in mass casualties.

Despite the evident appeal of such weapons as instruments of terror, terrorists have seldom used them. Terrorists have only been able to use chemical and biological agents rarely with little effect. The obvious reason has been the technical hazards which are considerable both in terms of acquiring / manufacturing the agent or weapons and in disseminating it in the environment. The probability of use of these types of weapons and the impact in terms of destruction and casualties they may cause are explained below:

**Figure 3.1 Potential Probability Vs Impact**

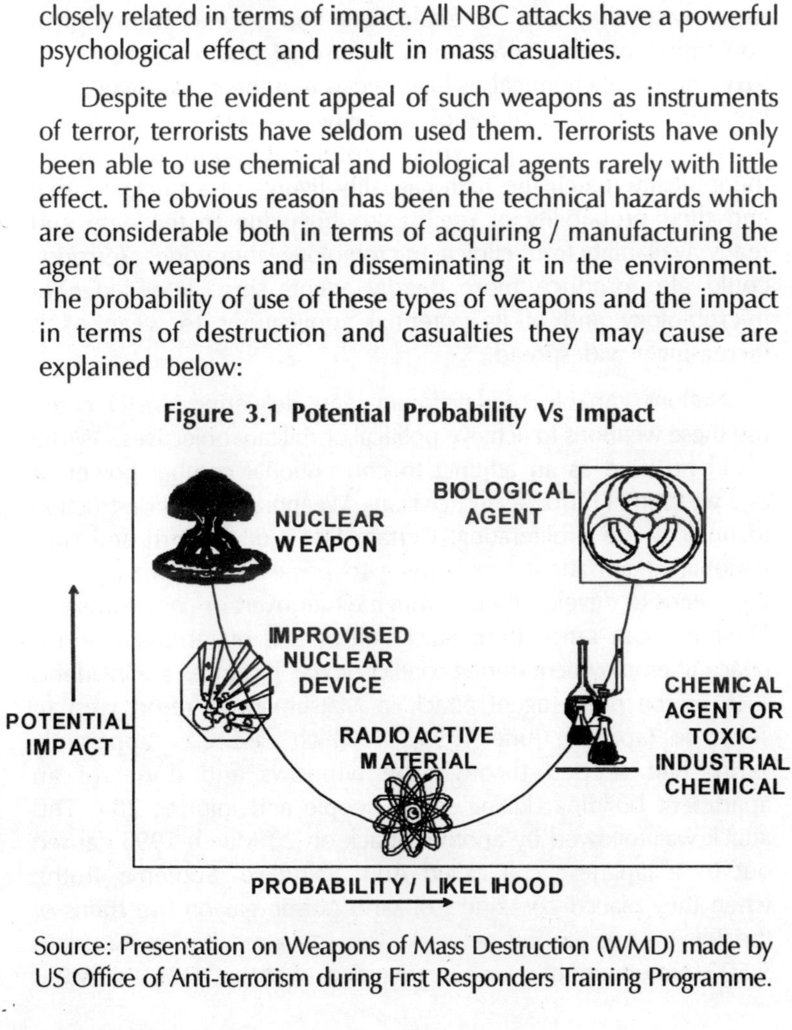


Source: Presentation on Weapons of Mass Destruction (WMD) made by US Office of Anti-terrorism during First Responders Training Programme.

The likelihood of use of nuclear device by the terrorist is negligible for now because of the technological and technical difficulties and the sophistication required to produce a nuclear bomb. The technical hurdles would be considerable; acquiring the weapon would present one set of difficulties, exploding it would present another. However, if the terrorists could lay their hands on radioactive material which can be found at a number of diverse facilities, including hospitals and industrial plants and in waste from nuclear plants, then they could detonate the same with the help of explosives and spread radiological contamination. The probability of use of chemical agents or toxic industrial chemicals is high because of their easy availability but their impact in terms of casualities is much lower. At the other end of the spectrum, biological agent use would theoretically match the high casuality figures of a nuclear strike and their probability of use is also high due to the easy and ready availability from clinical microbiology laboratories. Terrorists could also produce more deadly agents since knowledge of microbiology and of its potential applications to weapons is increasingly widespread.

Nations capable of developing and delivering WMD could use these weapons to achieve political or military objectives. WMD could be used as an adjunct to conventional combat power or as a weapon of terror against civilians. Weapons of mass destruction technology are proliferating. Certain States of concern and sub-national organisations are coming to possess these weapons or the means to develop them through either overt or covert transfer. Their use can range from blackmail or acts of terrorism during peace to employment during conflict or war. Instances are incidents such as the nerve agent attack in Matsumoto, a resort west of Tokyo in Japan in June 1994, in which case Gas (apparently sarin) had seeped through the windows and doors of an apartment building, killing seven people and injuring 284. This attack was followed by another attack on 20 March 1995 carried out by a Japanese cult called Aum Shinrikyo (Supreme Truth), when they placed containers of sarin poison gas on five trains of the Tokyo underground network. The attack resulted in 12 dead, 5,500 injured and of course, a great deal of chaos.[4] Police searches and investigations showed that the Aum specialists had experimented with a variety of chemical and biological agents and had dumped tons of dangerous chemicals in their warehouses. The underground attack left far fewer victims than could have been expected, partly because the chemical substance sarin used was not pure and because the means of distribution, polyethylene bags that had been punctured, were primitive. There was also a successful small scale attack in Oregon, USA where members of a religious cult (former members of Rajneesh's group) carried out the biological poisoning of a restaurant salad bar with salmonella typhimurium in 1984 resulting in poisoning of 750 people. By carrying out such acts, terrorists have demonstrated a knowledge of and willingness to use chemical and biological weapons to further their aims.

There are large number of factors, which have contributed to the growth of WMD terrorism. The growing incidence, sophistication and lethality of conventional terrorism have increased its shock value. Today, a large number of insurgencies /terrorist groups have state support and sponsorship. The threat of WMD weapons falling into the hands of terrorists has increased due to availability of nuclear weapons in unstable areas like the post-Soviet Union Russia and weak and unstable states like Pakistan. The WMD technologies are proliferating leading to growth of black and grey markets in WMD terror weapons.[5]

The technology for production of chemical and biological agents is proliferating.There are a large number of countries, which have matured chemical and biological weapons program despite the international treaties, that have banned their production and stockpiling. The possibility of terrorists acquiring these agents from the state favourable to their cause exists. Terrorists also might be able to steal chemical or biological agents from the state stockpiles with or without the connivance of the personnel guarding the facility. There have been attempts made by terrorists groups to buy the expertise from disgruntled scientists or the one who sympathise with their cause. Spearheading this trend is Osama bin Laden's al Qaeda outfit. This Islamic fanatical terrorist group, motivated by blind anger and rage, trying to avenge real or perceived slights and injustices, has compiled encyclopedias of jihad. It unsuccessfully sought weapons and radioactive dispersal devices. A senior Pakistani nuclear scientist Sultan Basiruddin Mahmood, who worked in the Pakistani nuclear programme since its inception in 1972, had been an admirer of Osama bin Laden and supporter of Taliban. He has claimed that only his knowledge about Pakistan's nuclear programme was a secret not to be revealed; the expertise he had acquired on enriching uranium or producing weapon grade plutonium, however, could be shared with others by him.[6]

A new class of WMD emerged with the employment of hijacked aircraft as flying missiles on 11 September 2001. The 11 September attacks on the USA, resulted in over 3,000 casualties. It may well be that terrorists will increasingly convert any means which are commonly available to cause heavy casualties without having to go through the rigorous process of acquisition of technological capabilities as envisaged for the use of chemical, biological or nuclear weapons. In future, technologically oriented terrorist groups could use explosive laden Unmanned Aerial Vehicles (UAVs) at selected high value targets with precision. This has posed a dilemma of categorising WMD terrorism. Is it only WMD terrorism using chemical, biological and nuclear weapons or does it include the large number of casualties suffered by states such as India in the past 57 years of independence, including the ongoing crisis in Kashmir where thousands of people have been killed? The scope of study has, however, been restricted to the use of WMD by non-state actors and terrorist organisations creating an asymmetrical threat for nations such as India.

## Nuclear Weapons

The great fear of the post-Second World War years has been the chance of use of nuclear weapons. Fears were further fueled by accidents or near accidents at nuclear reactors, culminating with the meltdown at Chernobyl and by the emergence of black market in nuclear materials.

***Characteristics:*** Nuclear weapons are similar to those of more conventional ones insofar as their destructive action is due mainly to blast and shock. On the other hand, there are several basic differences between nuclear and high explosive weapons. In the first place, nuclear explosions can be many thousand times more powerful than the largest conventional detonations. Second, for the release of given amount of energy, the mass of a nuclear explosive would be much less than that of a conventional high explosive. Third, the temperatures reached in nuclear explosion are much higher. In addition, a fairly large amount of energy is released in the form of light and heat referred to as thermal radiations.

The principal effects of nuclear weapons are blast, thermal radiation (heat), nuclear radiation and electromagnetic pulse (EMP). These effects are dependent on the size or yield of the weapons as expressed in kilotons, the physical design of the weapon, and the method of deployment.

**Table 3.2 Nuclear Weapons Effects**

| Effect | Percent of Nuclear | Products |
|---|---|---|
| Blast | 50% | Shockwave |
| Thermal Radiation | 35% | Heat, Fire, Burns |
| Residual Radiation | 10% | Fallout, Neutrons included Gamma activity |
| Initial Radiation | 4% | Casualties |
| Electromagnetic Pulse (EMP) | 1% | Electronic Interference |

Source: JCS J-3 *Weapons of Mass Destruction (WMD) Handbook,* Washington, DC (U.S. Department of Defense, February 2001), p.27.

A nuclear explosion results in four types of ionizing radiations: neutron, gamma, beta and alpha. Neutron and gamma radiation are present in the initial burst, while alpha, beta, and gamma rays comprise the residual radiation.

Electromagnetic Pulse (EMP) is the high energy, short duration pulse (similar in some aspects to a bolt of lightning) generated by a nuclear explosion. EMP can produce a current in any electrical conductor and temporarily disrupt or damage components not properly protected. Devices affected include radios, computers, telephones and other circuitry. Both EMP and transient radiation effects on electronic (TREE) must be taken into consideration in any situation involving the threat of nuclear weapons.

***Method of Dissemination:*** Any device designed to produce a nuclear yield can trigger a nuclear incident. The device may be a military grade nuclear weapon, or an Improvised Nuclear Device (IND). A nuclear device utilises the fission of nuclear material (U-235, Pu-239 etc.) to create blast and thermal effects and prompt exposure to gamma and neutron radiation. The fallout is composed of non-fissioned nuclear material, fission products and activation products. These materials contaminate very large regions downwind from the area impacted by the prompt weapon effects. Fallout creates very high external exposure hazards from both beta and gamma radiations and internal exposure hazards from inhalation or ingestion of the fallout.

## *Nuclear Terror*

Only the most sophisticated terrorist groups would likely to be considered as manufacturing their own nuclear weapons. For these groups the bending constraints would probably be the acquisition of fissile materials, which are much more highly protected than non-fissile radioactive isotopes. Mixed isotope plutonium (reactor grade material) can be used in nuclear weapons; such a device would be less efficient and might have a less predictable yield. Another alternative would be a radiological weapon that employed conventional explosives or other means to scatter radioactive material. Such weapons would not produce nuclear yield, however it could spread contamination. Such a weapon would have enormous potential for intimidation. Targeting a nuclear reactor in an opponent's territory to produce an accident, releasing nuclear material would be another viable option for the terrorists.

Fears were further fueled by accidents or near accidents at nuclear reactors, culminating with the meltdown at Chernobyl in USSR and by emergence of black market in nuclear materials. There is good cause to fear because proliferation continues and successful work is being done on nuclear devices in many states, some of which either openly or scrumptiously support terrorist groups.

There have been a few minor acts of nuclear sabotage, by radical ecologists albeit in minor scale during the 1970s and 80s. The first one occurred in 1973, when a commando team from a left wing Argentinean group, the ERP, entered the construction site of the Atucha Atomic Power Station north of Buenos Aires. In 1976, bombs were thrown at an atomic power plant in Brittany, France, but the nuclear reactor was not damaged. In 1982, the terrorist wing of African National Congress (ANC) sabotaged two South African nuclear plants, causing substantial damage. Both the nuclear reactors were damaged, but since they were not operational at that time there was no emission of radiation.[7] The bombing or sabotage of a civilian reactor might cause relatively less direct damage, provided it is located far from major cities, the ecological damage, on the other hand, would be great and lasting. But in view of the fact that so many reactors now exist around the globe, that many of them are located in or near major cities and that radiological materials are also stored in laboratories and hospitals, even a small amount of radiological contamination could have major consequences.

Although the probability of nuclear terrorism is very low, its horrendous consequences demand continuing vigilance. Nuclear power plants cannot withstand attack by a large jumbo jet full of fuel. The Director General of International Atomic Agency (IAEA) has warned, "The tragic terrorist attacks on the United States were a wakeup call to us all; we cannot be complacent".[8] There is therefore an urgent need to implement security upgrades at all our nuclear facilities.

Nuclear terrorism usually subsumes a wide range of malevolent activities in order to blackmail the targeted state to accede to the demands of the terrorists threatened or actual use of a nuclear device, targeting nuclear facilities like power stations, breeder reactors, reprocessing and enrichment plants or cooling ponds of spent nuclear fuel. Nuclear installations contain enormous inventories of radioactive materials and are, therefore, attractive targets for blackmail. The fear of radioactivity adds new dimension to terror. Till date more than 50 attempts to exhort money with nuclear threats have occurred in US and some of them were frighteningly credible. There also have been 643 incidents of illicit nuclear trafficking between January 1991 and December 2001, but only 48 cases of those were of high proliferation concern.[9] Terrorists may resort to nuclear blackmail but according to some commentators, no terrorist group has even seriously attempted to seize possession of a nuclear weapon. However, in the recent years, the so-called new terrorists have emerged who despise the entire social order and would like to give a push to history. They tend to be motivated by religious fanaticism or millenarian prophecy. Violence assumes a transcendental dimension and is geared to take revenge for real or perceived injustices.

Presently the amount of nuclear material and strong radiation sources not classified as nuclear material all over the world is as under:[10]

(a) 438 nuclear power plants with, typically, several thousand fuel rods each.

(b) About 650 research reactors with typically several tons of fuel elements each.

(c) 250 nuclear fuel-cycle plants with an unknown amount of material.

(d) More than 10,000 teletherapy sources with one Cobalt-60 capsule each; several hundred teletherapy sources with one Caesium-137 capsule each.

(e) About 300 industrial irradiators with numerous rods each.

(f) Several tons of thousands of industrial radiography sources (80% Iridium-192, remainder Cobalt-60, Selenium-75, Ytterbium-169).

(g) About 10,000 Iridium-92 industrial radiography sources added every year; Over 1,000 Cobalt-60 sources now in circulation.

(h) About 1,000 sources of Selenium-75 and Ytterbium-169 supplied annually.

The numbers of incidents of illicit trafficking in Nuclear and Radioactive Materials the world over are less in comparison to the figures of early 1990s. The details are shown below.[11]

**Fig 3.2 Confirmed Incidents with Nuclear Material 1993-2001**

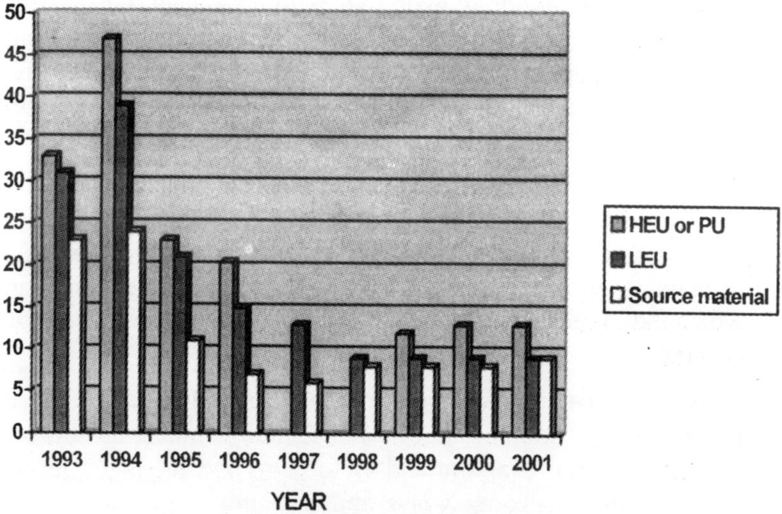


**Fig 3.3 Confirmed incidents with Nuclear Material and Radioactive Sources 1993-2001**

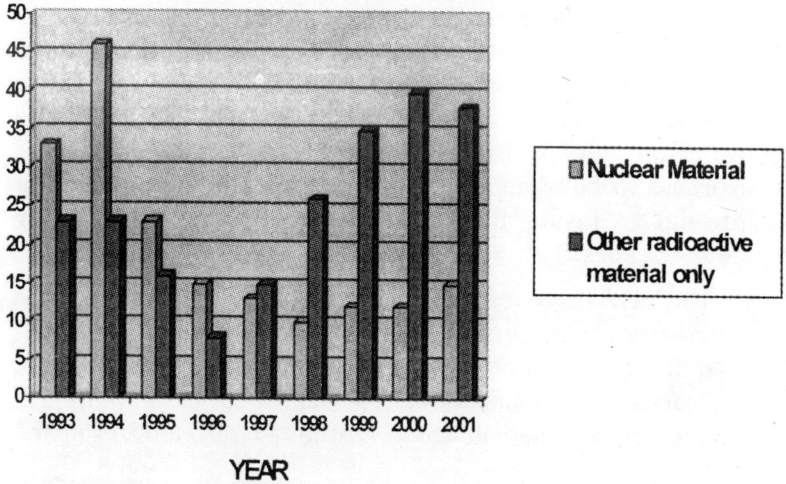


(*Source:* IAEA Databank).

## Radiological Weapons

Terrorists might not be able to procure sufficient weapons grade plutonium nor have the expertise to build a functional nuclear bomb, but they could acquire enough radioactive material to kill civilians and create panic with hybrid weapons. These 'dirty bombs' or radiological weapons can be made by spiking a conventional explosive with low-level radioactive material obtained from the black market or even stolen from hospitals, dumps or factories. These weapons do not pack the punch of nuclear weapons, but the psychological toll and contamination they wreak would be considerable.

***Agent Characteristics:*** A radiological weapon causes injury by exposing people to alpha, beta and gamma particles. This weapon unlike a nuclear weapon, does not release radioactivity in massive burst of energy. Rather it mimics a toxic chemical in which the cumulative dose of radiation eventually proves harmful or fatal.

Radioactive materials can include those commonly found in nuclear plants (Iodine, Tritium and Cesium), industrial sources (Cobalt, Cesium, Uranium and Plutonium) and medical sources (Phosphorus, Thallium, X-rays and Sodium).

***Methods of Dissemination:*** A Radiological Dispersion Device (RDD) uses any mechanism to distribute radioactive material over the target area. The actual means of dispersal could include conventional explosives, aerosol sprays or hand distribution. The hazard is dependent on the type of radioactive material involved, its physical form and the nature of the dispersion mechanism. Hazards created include external exposure to beta/gamma radiation, gamma radiation alone or internal exposure from alpha, beta or gamma emitting radioneuclides that are ingested or inhaled.

The advantage of radiological weapons is that nuclear technology is not required to create them. Anyone with access to radioactive materials can build a radiological weapon. Radioactive contaminants can be delivered by a variety of means including human agent, missile, aircraft, artillery or the destruction of a facility containing radioactive materials. A terrorist could either incorporate readily available radiological materials into an explosive device or use a large explosive to damage a radiological site such as a nuclear power plant. It is often difficult to determine that a radioactive material is involved therefore, necessary detection equipment should be put to use at all explosive incident sites.

## Radiological Terror

Common radioactive materials, such a commercial radioactive sources used in medicine, industry and scientific research could fuel radiological terrorism by building and using radiological dispersal devices (RDDs) – one type of which is popularly known as a "dirty bomb". RDDs are not nuclear weapons and generally would not cause massive destruction, but they could spread radioactive particulates over wide areas. Although few people if any, would die shortly after exposure to the ionizing radiation from a typical RDD, many could panic and become terrorised because of the fear of radioactivity.

Factors determining the security risk of a type of radioactive source include prevalence of use, radioactive content, portability and dispersibility. Generally, the more prevalent, a radioactive; portable and dispersible source is, the higher security risk it presents. For example, cesium chloride containing relatively large amounts of radioactive cesium – 137 and consisting of an easily dispersible powder would definitely be categorized as a high security risk compound. If this material were also housed inside a portable container, a thief or a terrorist could readily seize and transport the radioactive source if adequate security measures are absent.

Although dozens of different radioisotopes are employed in radioactive sources throughout the world, only about eight radioisotopes have characteristics that raise the sources containing them to the highest security risk level. Each chemical element, such as cesium, comes in different form called isotopes that have the same chemical properties, but different nuclear characteristics. Unstable isotopes, termed radioisotopes, try to transform into stable isotope states by emitting radiation.

The radioisotopes of highest security concerns include the reactor produced americium-241, californium-252, cesium-137, cobalt-60, iridium-192, plutonium-238 and strontium-90 as well as the naturally occurring radium-226. The half-life (the time required for half the radioactive material to decay) of most of these radioisotopes is years to decades in length. After seven half-lives, the radioactive substance has decayed to less than one percent of its initial amount. Therefore, most of the high risks radioactive sources will emit the majority of their radioactivity during a time period covering substantial parts of, or all of a typical human lifespan. This fact explains part of the reason why radiological dispersal devices using these materials raise the risk level to human health.[12]

Other risk factors to human health stem from the ionizing radiation emitted by those eight radioisotopes. Four of the isotopes (americium-241, californium-252, radium-226 and plutonium-238) primarily release alpha particle radiation and would mainly pose internal health hazards via ingestion or inhalation because the dead layer of skin on a human body stops alpha particles. Three of the other radioisotopes censium-137, cobalt-60, and iridium-192) result in the emission of high-energy gamma radiations and would pose both external and internal health hazards because radiations can easily pass through the body's dead layer of skin. Because strontium-90, the remaining radioisotopes on this high security list, emits high-energy beta particles, it could pose an external health hazard in absence of shielding. But it is primarily an internal hazard because if ingested it concentrates in bone.

The first threat of an RDD attack occurred on 23 November 1995 when Chechen guerrilla leader, Shamyl Basayev, informed the Russian TV Network NTV that four cases of radioactive Cesium had been hidden around Moscow. A 32-Kg case was discovered in Ismailovo Park, Moscow. Though it contained Cesium 137 used in X Ray equipment, the possibility of terrorists using an RDD was established.[13]

## Biological Weapons

Biological warfare comprises use of agents such as viruses, bacteria, and fungi, generally classified as infectious microbes, which can affect humans, animals and plant life alike. Characteristics of biological agents include minimum contagiousness to prevent infecting the population of the home country; hence diseases such as small pox are best avoided while anthrax is ideal. Spread of diseases, including plague, to a target country through innocent travellers easily providing anonymity, till the attacker decides to reveal his identity, is an advantage of the use of biological weapons.

The history of biological warfare goes back much further than the use of poison gases, but the biological agents that threaten us today are of recent origin. The great plague of the fourteenth century, which is said to have killed about one-third of the population of Europe, was allegedly spread by Tartars besieging the fortress of Caffa in the Crimea. In 1346, the Tartars catapulted plague-infected corpses over the walls into the fortress; according to reports, ships from Genoa carried the disease to northern Italy and the rest of Europe. In the 1500's, the Spanish conquistadors used biological (small pox) warfare against the natives of the Americas. In 1710, the Russian troops reportedly used plague infected corpses against the Swedes and in 1767, the English during the French and Indian war in North America provided blankets to the Indians that had been infected with small pox.

Sporadic uses of biological agents were reported during the First World War. In World War 1, the Germans are reported to have infected Romanian cavalry horses, as well as livestock in the United States, being dispatched for use by the Allies with glanders, with a bacterial disease.[14] Japan and Britain started Biological weapons programme in the 1930's and the Soviet Union and the United States followed suit. In the mid-1930s, the Japanese Army established a special biological warfare unit called '731'.[15] Many biological agents were produced in the laboratories of the unit, including plague, small pox, typhus and gas gangrene. These agents were tested on the Chinese prisoners of war and civilians and about ten thousand people were killed.

Ambiguity in the content, goals and potential applications of biological warfare research has been recognised by the world at least since the early 1940's. A report resulting from a review of the United States Biological Warfare Programme in 1949 acknowledged the overlap of the offensive and defensive preparations. Noting that, *"the offensive employment of biological weapons is predicated upon the ability to immunise own troops, those of our allies or other personnel likely to come within range of infection by our own biological warfare weapons".* The report went on to comment on the duality of biological warfare research, *"information obtained from research on the defensive aspects of Biological warfare is, in the greater part, applicable to offensive problems as well... Defensive warfare research has many peace time applications, particularly in the fields of medicine and public health".*[16]

Biological research currently supported by the governments world over involves considerable ambiguity. The classification and analysis of the projects under study pose two general problems. First is the ambiguity of the content of the research and secondly, that many of the pathogens under study as potential biological warfare agents are also endemic in various parts of the world.

***Characteristics:*** Biological warfare (BW) is the employment of pathogens or toxins to produce casualties in humans or animals and damage to plants or material. How do pathogens and toxins differ? Pathogens are living organism, while toxins are the inert by-products or waste materials of living organisms. Pathogens are disease-producing microorganisms, such as bacteria, viruses, rickettsia, or fungi. They can be naturally occurring, or altered by mutation or genetic engineering. Toxins, on the other hand, are poisons produced by the metabolic activities of living organisms. The most common biological agents include anthrax, botulinum toxin, cholera, smallpox, tularemia, Q fever, ricin, and plague.

BW represents a less sophisticated and often cheaper alternative to nuclear or chemical weapons. BW agents are inherently more toxic then chemical agents on a weight-for-weight basis and can provide broader coverage per pound of payload. Moreover, they are potentially more effective because most are naturally occurring organisms such as bacteria and viruses, which are self-replicating and have specific physiologically targeted effects, whereas chemical agents are manufactured chemicals that disrupt physiological pathways in a general way. They are effective in areas with low hygiene and poor medical facilities.

BW agents can be produced with the same equipment as used for the manufacture of medical vaccine and can be employed directly or indirectly against personnel, plants, animals, or material. Foods, especially uncooked food as in a salad bar, water supplies, or facilities can be rendered unsafe or unfit for use or consumption through contamination. People can be infected either directly by the employment of a BW agent or indirectly through secondary contamination from an exposed individual.

***Dissemination:*** BW agents (whether pathogen or toxin) are most likely to be disseminated using the following methods:

(a) *Line Source:* This technique is most effective using a dispersal means (a truck or air sprayer like a crop duster) moving perpendicular to the wind during an inversion (in which air temperature increases with altitude, holding surface air and pollutants down; incursions normally occur at dawn, dusk, or night).

(b) *Point Source.* This technique involves delivering a specific quantity of agent to a location or individual, or the releasing of bulk quantities of agent in a saturation mode. An example of this is the mailing of anthrax-contaminated letters to targeted individuals. Or, a lone individual at a sporting event releasing powdered BW agent for the wind to disseminate throughout the stadium. BW agents could also be introduced into the ventilation system of a targeted building or used to contaminate food or water.

## Biological Terror

Biological warfare agents are living organisms or the materials derived from them (endotoxins or extoxins) that cause disease in or harm humans. A biological agent has a latent period between infection and onset of disease. The onset of symptoms varies with the microorganism and may vary from days to weeks. There typically will be no characteristic signature because biological agents are usually odorless and tasteless.

Biological agents pose a significant threat because their use is even more difficult to recognise than the use of chemical agents. The presence of symptoms may well be confused with a naturally occurring case or outbreak of disease. Many of the initial symptoms may be common to several other types of disease, which further complicates recognition, identification, and treatment. The ease with which people can travel throughout the world today presents the possibility of an infected individual carrying the infection home before becoming symptomatic. The recent outbreak of the Ebola virus in Zaire is an example of how a dangerous disease can quickly and easily spread. Fortunately, the disease was confined to the local area.

Biological organisms mutate and often become even more deadly and resistant to available drugs. Biological organisms can be obtained from secretions or tissues from sick or dead individuals. Biological organisms can be cultured or grown relatively easily. Controls on biological materials are not as stringent as those for chemical agents. Small quantities can easily be transported undetected.

The most alarming aspect of biological agent use is the fact that an infected person may unknowingly transmit the disease to others. The contagious nature of many biological agents creates a situation in which the disease can spread far beyond the incident site. Unlike chemical agents, once a biological agent organism finds a suitable host, it can reproduce and continue to spread the disease.

Biological agents have a few well-known advantages over other agents. They are easy to produce and difficult to detect. They are cheap and can cause human fatalities as well as economic damage by ruining crops and live stocks. The value of biological warfare to the terrorist does not lie so much in the extent of the damage that would be inflicted but rather in the panic and psychological impact even a small success would be likely to engender. Just as one of the chief military virtues of strategic biological warfare was the 'anxiety factor' and the public's fear of the disease, use of biological warfare agent would have a propaganda value far beyond the likely casualty level. Plague would be especially effective in this regard even without epidemic spread because of its historical association with the great pandemics of the middle Ages. Even a small biological attack in a large city would cause the public to fear further outrages anywhere at any time and exploit the psychological value of BWs 'invisibility'. The attractions of biological weapons are obvious: easy access, low cost, toxicity and the panic they can cause. But there are drawbacks of various kinds that explain why almost no successful attacks have occurred. While explosive or nuclear devices or even chemical agents, however horrific, affect a definite space, biological agents are unpredictable; they can easily get out of control, backfire or have no effect at all. They constitute a high risk to the attackers. Biological agents with some notable exceptions are affected by changes in heat or cold and directions of wind. They have limited life span and their means of delivery are complicated. Dispersing the agent as a vapour or via an aerosol system within a closed space, for instance through the air conditioning system of a big building or in a subway would appear to offer better chances of success, but it is by no means foolproof.

There are well known defences against biological weapons, ranging from immunisation to gas mask and other protective clothing to decontamination process. Antibiotics are effective against many bacteria if applied soon after infection; and there are various antifungal medicines. However, in biological warfare supply side controls face even more daunting prospects in the realm of biological weapons. Virtually all the equipment, technology and materials needed for biological warfare agent research, development and production are dual use. This makes offensive programme relatively easy to disguise within the larger body of legitimate commercial activity, as no specialised facilities are required and any country with the political will and a competent scientific base can produce toxins or infectious agents.

Although, only three or four nations were thought to have offensive biological weapon programme when the Biological Weapon Convention (BWC) entered into force in 1975. The intelligence community currently assesses that perhaps a dozen states or so maintain offensive programmes. They have also warned that "credible biological warfare capabilities are becoming more advanced", a trend that may enhance the prospects of biological weapons use in the years ahead.[17] The lasting demand for biological weapons, the relative ease of concealing any offensive effort, the growing availability of weapons related technologies that could significantly alter the threat environment, all suggest that determined states as well as particular sub nationalactors, especially those supported by states face few real constraints in establishing, developing or improving offensive programme with a national decision to do so.

While a large number of microbes have been used as bio weapons, a recent report by Center for Diseases Control, Atlanta, USA has divided biological agents into three categories according to the level of risk they present to the public.

(a) *Category A*: Category A includes the highest priority agents, which are microbes that are easily disseminated or transmitted person to person with major public health impact, cause high mortality, public panic and social disruption and require special action for public preparedness. Included in this category are Variola Major, Bacillus anthracis, Yersinia pestis, clostridium botulinum toxin, Ebola virus and Francisella tularensis, Marburg hemorrhagic fever, Lassa virus and Junin and related viruses.

(b) *Category B*: Agents included in this category are those that are moderately easy to disseminate, cause lower mortality but significantly morbidity and require specific enhancements of diagnostic capacity and, enhanced disease surveillance. Agents included are coxiella burnetti, Brucella species, Meliodosis, VEE, Typhus fever, toxic syndrome, food safety threats and water safety threats due to Salmonella species, shigella dysteneriae, E. coli and Vibrio Cholerae.

(c) *Category C*: Agents include emerging pathogens that could be engineered for mass dissemination in the future due to their availability, ease of production and potential for high morbidity and mortality. Emerging threat agents are Nipah Virus, Hanta virus and SARS.

Accidents have occurred in germ-warfare installations, the most famous of which was the anthrax epidemic at Sverdlovsk in the Urals in USSR in April 1979. According to western intelligence, a huge airborne (aerosol) release of anthrax spores used for bacteriological warfare resulted in many fatalities.[18]

## Chemical Weapons

The third of the trinity of weapons of mass destruction is chemicals being employed as warfare agents. Nations across the world possess an extensive arsenal of chemical weapons, delivery systems, protective equipment and an extensive chemical force structure which exceeds reasonable requirements. Most of the countries world over has 'Chemical Corps' as part of their armies.

The Geneva Protocol of 1925 prohibited the use in war of asphyxiating, poisonous or other gases and of bacteriological (biological) methods of warfare. The chemical weapons convention of 1993 prohibits development, production, stockpiling, acquisition or retention of chemical weapons as well as their use. This convention has come into force in 1997.

Despite widespread collective aversion towards chemical weapons, this means of warfare continues to be of major concern especially from terrorists, from states of concern that support them and from fractious ethnic and religious groups.

Chemical Warfare Agents have been defined in a report authorised by the United Nations General Assembly thus: "Chemical agents of warfare are taken to be chemical substances, whether gaseous, liquid or solid, which might be employed because of their direct toxic effect on man, animals and plants.[19]

A chemical agent can be defined as a chemical substance intended to kill, seriously injure, or incapacitate through its physiological effects. A chemical agent attacks bodily organs in such a way that it prevents those organs from functioning normally. The results are usually disabling or even fatal. Not included in this definition are riot control agents such as tear gas, smoke and flame materials. Chemical warfare (CW) agents are separated into groups according to the potential severity of their effects; lethal, blister, or incapacitating. Lethal chemical agents are those designed to cause death among exposed personnel. They include nerve, choking, and blood agents.

Blister agents damage any tissue, which they come in contact with. They could be lethal under certain conditions, but skin, eye, and respiratory damage are their main casualty-producing effects. Blister agents could also be employed to contaminate terrain, facilities, or material. Incapacitating agents are chemicals that cause physiological or mental effects that lead to temporary disability. Unlike riot control agents such as tear gas, which have effects lasting only a few minutes, incapacitating agents produce effects that can last from hours to days.

Chemical agents are classified as non-persistent, as in the case of the G-series of nerve agents, or persistent, as with the nerve agent VX and blister agents. The persistency of a chemical agent refers to the duration of its effectiveness under certain conditions after its dissemination. Generally speaking, persistent agents can present a contact hazard as well as an inhalation hazard. Non-persistent agents present only an inhalation hazard. In order to achieve good ground coverage when dispersed from a high altitude with persistent CW agents, the dispersed droplets must be sufficiently large to ensure that they fall within the target area and do not get transported elsewhere by the wind.

This can be achieved by dissolving polymers (e.g. polystyrene or rubber products in the CW agent to make the product highly viscous or thickened. The result is that the persistence and adhesive ability is increased. This complicates decontamination.

***The Nerve Agents:*** Amongst lethal chemical agents, nerve agents have had an entirely dominant role since 1945. Nerve agents disrupt the transmission of nerve impulses within the nervous system. Nerve agents belong chemically to the group of oregano-phosphorus compounds that include many commercially available pesticides. Nerve agents are stable and easily dispersed, highly toxic and have rapid effects whether they are absorbed through the skin, eyes, or inhaled. All nerve agents in their pure state are colorless liquids. Readily available industrial chemical processes can manufacture them. The ingredients, or precursors, of nerve agents, although controlled by international treaties such as the Chemical Weapons Convention (CWC), are obtainable. The most important nerve agents are:

(a) Tabun, GA.

(b) Sarin, GB.

(c) Soman, GD.

(d) VX.

***Physical and Chemical Properties***: Nerve agents are characterized by their extreme toxicity and rapid effect. A nerve agent, either as a gas, aerosol or liquid, enters the body through inhalation or by absorption through the skin. Poisoning may also occur through consumption of liquids or foods contaminated with nerve agents. The manner in which it enters the body impacts on the time required for the nerve agent to start having effect. It also influences the symptoms developed and, to some extent, the sequence of the various symptoms.

Generally, the poisoning works faster when the agent is absorbed through the respiratory system than via other routes. This is because the lungs contain numerous blood vessels and the inhaled nerve agent can, therefore, rapidly diffuse into the blood circulation and thus reach the target organs.

Amongst these organs, the respiratory system is one of the most important. If a person is exposed to a high concentration of nerve agent, death may occur within a matter of minutes. Poisoning takes longer when the nerve agent is absorbed through the skin. Nerve agents are fat-soluble and can penetrate the outer layers of the skin. However, it takes some time before the poison reaches the deeper blood vessels. Consequently, the first symptoms do not occur until 20-30 minutes after the initial exposure, but the poisoning process may be rapid if the total dose of nerve agent is high.

The toxic effect of nerve agents depends on its binding to the enzyme, acetyl cholinesterase, and thereby inhibiting the enzyme's normal function in the cholinergic nervous system.

***Symptoms:*** When exposed to a low dose of nerve agent, characteristic symptoms include increased production of saliva, a running nose and a feeling of pressure on the chest. The pupil of the eye becomes contracted (miosis) which impairs night-vision. The accommodation capacity of the eye is also reduced so that short-range vision deteriorates and the victim feels pain when he tries to focus on an object nearby. This is accompanied by headache. More unspecific symptoms are fatigue, slurred speech, hallucinations, and nausea.

Exposure to a higher dose leads to a more dramatic development and symptoms are more pronounced. Bronchoconstriction and secretion of mucous in the respiratory system leads to difficulty in breathing and coughing. Discomfort in the gastrointestinal tract may develop into cramps and vomiting. Involuntary urination and defecation may also form part of the picture. There is a large discharge of saliva and the victim may experience running eyes and sweating. Symptoms from the skeletal muscles are very typical. If the poisoning is moderate, this may express itself as muscular weakness, local tremors, or convulsions.

When exposed to a large dose of nerve agent, the muscular symptoms are more pronounced. The victim may suffer convulsions and lose consciousness. To some extent, the poisoning process may be so rapid that earlier mentioned symptoms may never have time to develop. Muscular paralysis caused by nerve agents also affects the respiratory centre of the central nervous system. The combination of these two effects is the direct cause of death. Consequently, death caused by nerve agents is a kind of death by suffocation.

The toxic effect depends on both the concentration of nerve agent in the air inhaled (C) and time of exposure (t). In extremely high concentrations there is a simple relationship, Ct, which gives a certain toxic effect. Inhalation of sarin vapors with a concentration of 100 mg/m3 for one minute gives the same result as inhalation of 50 gm/m3 for two minutes. However, at low concentrations this relationship does not apply, since the human body is capable of some degree of detoxification. In order to obtain a corresponding effect, it is then necessary to have relatively longer periods of exposure.

***Blood Agents:*** Most blood agents are non-persistent CW agents, which are primarily absorbed into the body by breathing. These agents can be dispersed by artillery, mortar, rocket, aircraft spray, or bomb. There are two primary blood agents — hydrogen cyanide (AC) and cyanongen chloride (CK). It has been reported that hydrogen cyanide was used by Iraq in the war against Iran and against the Kurds in northern Iraq during the 1980s. Hydrogen cyanide has high toxicity and can, in sufficient concentrations, rapidly lead to death. During the Second World War, a form of hydrogen cyanide (Zyklon B) was used in the Nazi gas chambers. At room temperature, hydrogen cyanide is a colorless liquid.

Both gaseous and liquid hydrogen cyanide can also enter the body through the skin. Its high volatility probably makes hydrogen cyanide difficult to use in warfare since there are problems in achieving sufficiently high concentrations outdoors. On the other hand, the concentration of hydrogen cyanide may rapidly reach lethal levels if it is released in confined spaces.

Symptoms of cyanide poisoning vary and depend on, for example, route of poisoning, total dose and the exposure time. If hydrogen cyanide has been inhaled, the initial symptoms are restlessness and increased respiratory rate. Other early symptoms are giddiness, headache, palpitations and respiratory difficulty. Vomiting, convulsions, respiratory failure and unconsciousness later follow these. If the poisoning occurs rapidly, e.g. as a result of extremely high concentrations in the air, there is no time for symptoms to develop and exposed persons may then suddenly collapse and die. The treatment given to victims is based on encouraging and speeding-up the body's own ability to excrete cyanide and to bind cyanide in the blood.

***Blister Agents:*** Blister agents received their name because of the similarity of the wounds they caused to burns and blisters. However, since blister agents (also called "vesicants") also cause severe damage to the eyes, respiratory system and internal organs, they are also described as tissue-injuring agents. The effect of blister agents are delayed and the first symptoms do not occur until 2 to 24 hours after exposure. The severity of a blister agent burn is directly related to the concentration of the agent and duration of its contact with the skin. Most blister agents are relatively persistent and may be employed as a gas or liquid. Blister agents are readily absorbed by all parts of the body. Poisoning may also occur through consumption of liquids or foods, which have been contaminated. The most common cause of death as a result of blister agent poisoning is complications after lung injury caused by the inhalation of the agent. The majority of chronic and late effects from blister agent poisoning are also caused by lung injuries.

***Choking Agents:*** Choking Agents are non-persistent, lethal chemical agents designed to cause death to an exposed individual. These agents injure unprotected personnel mainly in the respiratory tract (nose, throat, and lungs). These agents will "choke" an unprotected person. Upon exposure, membranes swell and secrete fluid, the lungs fill with this fluid, and death results from lack of oxygen. The term for such a death is "dryland drowning". There are two choking agents: Phosgene (CG) and diphosgene (DP). CG is a chemical agent with short agent-cloud duration. It was used extensively during World War I. An estimated 80 Percent of World War chemical fatalities resulted from CG. CG is a colorless gas with an odour similar to new-mown hay, grass or green corn. It tends to hug the ground and linger for extended periods of time in trenches and low areas.

Diphosgene (DP) is also a colorless gas with an odour similar to CG. DP has a stronger tearing effect than CG and thus has less of a surmise value when used against personnel. Its symptoms and effects are similar to CG.

***Incapacitating Agents:*** This group of agents includes substances which, when administered in low doses, cause conditions similar to psychotic disorders or other symptoms emanating from the central nervous system (loss of feeling, paralysis, rigidity, etc.). The effects are transitory and cause inability to make decisions and incapacitation. During the 1950's, studies were made on substances such as glycolic acid esters (glycoalates). Particular interest was paid to 3-quinuclidiny lbenzilate, BZ. The effects of BZ cause poisoning at doses of 0.5-5 mg. Peripheral symptoms such as distended pupils, deteriorated short-distance vision, dry mouth and palpitations occur after about 30 minutes.

A serious effect of poisoning with BZ, as also with other atropine-like substances, is an increased body temperature. Deterioration in the level of consciousness, hallucinations and coma occur subsequently. Incapacitating after-effects may remain one to three weeks after the poisoning.

***Methods of Dissemination***: Bombs, spray tanks, missiles can be used to deliver chemical agents. Terrorists could also employ chemical agents covertly with improvised devices as was done by the Aum Shinrikyo cult. In the dispersal of CW agents, a mixture of liquid droplets and gas is generated. The weather, terrain, and buildings will all have an effect on the chemical agent once it has been employed.

***Influence of Weather***: A terrorist attack involving chemical weapons will cause injuries to unprotected people who are in or downwind of the target area. Strong wind, heavy rain or temperatures below freezing point may reduce effects. After the attack, the weather will be of great importance for the respiratory risks expected at different distances from the target. Similarly, weather conditions influence the effect of ground contamination. After an attack, the primary cloud will drift with the wind. Wind velocity will be decisive for how long it will take for the primary cloud to pass the given place. High wind velocity implies a short time of passage and thus fewer injuries to unprotected persons, whereas low wind velocity will lead to more injuries. Consequently, a weak wind may cause effects at greater distances than strong winds.

Wind velocity also influences how fast the primary cloud moves. If the wind is gentle, then there are better opportunities to provide warning in time. In case of very weak winds, however, the gas cloud will not move very far. In addition, the wind direction varies widely in such situations, which is why a circular area must be alerted in the event of an attack with CW agents. The concentration in the primary cloud may also decrease in cold weather and particularly if the temperature is lower than freezing. This depends on a smaller amount of CW agents evaporating during dispersal, which also implies that the share of ground contamination will be greater.

Precipitation also reduces the concentration, since some of the gas or aerosol is "washed" away by wet deposition. A major problem during the winter may be that contaminated snow on shoes and clothes is taken into shelters, vehicles or buildings. Once in the warmth, the CW agent will evaporate and may cause concentrations of gas. Light rain will cause ground contamination to be more dangerous since the pores in the soil become clogged and prevent the substance from penetrating down into the soil. Heavy rain, however, will flush off ground contamination whereas heavy snow will cover it. In both cases, the contact risk is decreased.

***Terrain and Buildings*:** Woodland and undulating terrain give shorter danger distances for the primary cloud since the wind will be exposed to greater turbulence. Woodland also absorbs a certain amount of gas and aerosol through dry deposition. In or close to the target area, however, woodland, depressions, pits and narrow streets may lengthen the effect of an attack. Gas and aerosol will be retained in these areas, particularly in situations of weak wind and stable stratification. The longest danger distances are obtained if the cloud passes over plains or lakes, or follows the contours of a valley.

The effect of a passing cloud of gas or aerosol will be delayed inside tents, buildings and vehicles. Owing to the lower air exchange in such places, it will take longer for the cloud to penetrate. A certain amount of the CW agent will be taken up and bound on walls and other surfaces, which also contributes to decreases in concentration. Consequently, it may be expected that there is a certain reduction in the effect of a passing cloud of gas. In ordinary buildings, the protection can be improved by closing doors and windows, turning off the ventilation and sealing all cracks with tape. It will, thus, take longer for the cloud of gas to penetrate a house, for example. For the same reason, the gas will remain longer in doors when the rest of the cloud has passed by. Consequently, it is of the greatest importance to obtain information on when it is possible to start airing the building again.

Ground configuration is also of importance for the contact risk in ground contamination. A dry, hard but porous surface, e.g., asphalt or concrete, will lead to fewer contact risks. On soft ground, e.g., grass, moss, sand or snow, it is easier to come into contact with CW agents that have penetrated the underlying surface. In dense woodland, the ground contamination is reduced and becomes uneven since the falling droplets are caught to some extent in the crowns of the trees. Terrain covered by bushes, on the other hand, may lead to major risks of contact.

## Chemical Terror

Chemical warfare envisages employment of chemical compounds, generally toxic in nature, in conflicts. The potent effects of chemicals, both lethal and non-lethal, on the human organism include paralysis of the nervous system, blinding albeit for short periods, loss of hearing, burns, difficulty in breathing and, at the least, nausea and vomiting. Use of chemical defoliants and herbicides in warfare has also been designated as chemical warfare. A 1969 report from the United Nations defines chemical warfare agents as "chemical substances, whether gaseous, liquid or solid, which might be employed because of their direct toxic effect on man, chemicals and plants". The Chemical Weapons Convention describes chemical weapons as "any chemical which through its chemical effect or living processes may cause death, temporary loss of performance, or permanent injury to people and animals."[20]

The idea of using gas against one's enemies occurred first to the Fenians in the 1870s, who intended to spray it in the House of Commons in London. There seems to have been similar plans during the Boer War and the Japanese War with the Russians in 1905, but these did not go beyond an amateurish experimental stage. Germans in Ypres first used gas on a massive scale in 1915. The substance used, chlorine gas, came as a total surprise to the

Allies. This attack resulted in five thousand dead, many injured and a German breakthrough. However Germans failed to follow up the initial panic. Seen in retrospect, this gas warfare was a failure at that time. There were stray cases of use of poisonous gases during the First World War, but with limited number of casualties. However, during the Second World War the poison gases were not used, since both sides possessed the capability to wage chemical warfare against the other.

Since World War II, chemical weapons have been reportedly used in Yemen in 1966-67 and by Iraq in the eight yearlong Iraq-Iran war. In each of these, initial use of chemicals was against troops, which lacked even the most limited protection against chemical attacks such as gas masks. Till September 2002, the most notorious attack involving chemical weapons against a civilian target remained the use of Sarin nerve agent in a Tokyo subway in March 1995, by the Japanese cult, Aum Shinrikyo.[21] This was followed by simultaneous release of the gas in five underground trains bound for central Tokyo. The incident resulted in the death of 12 people and 5,500 injured. An earlier attack on Matsumoto on 24 June 1994, in which seven people had died and a large number were injured, had gone relatively unnoticed. The cult wanted to destroy the world "outside", and wanted to follow it up by use of nuclear weapons to destroy life on Earth and become "Gods".[22] Aum Shinrikyo represented a unique group of terrorists of apocalyptic desires, revenge, criminal intent and material greed. Such groups are most prone to employ WMD, not merely to terrorise but also exterminate, to the extent possible, life on Earth itself.

Today in contrast to biological warfare, chemical warfare has already been waged on a massive scale. Today, we have a large number of poisonous chemical substances not in the form of gases but liquids, which can be dispensed in droplets. They can be divided according to their chemical composition. Choking agents in the nature of chlorine and phosgene which cause pulmonary edema were used in World War 1, along with blistering agents in the form of mustard gas which caused chemical burns and destroyed lung tissues. Hydrogen cyanide and Cyanogens chloride attack the respiratory system and result rapidly in coma and death. The nerve gases tabun, sarin, soman and VX affect the neuromuscular system, culminating in death.

Most of the chemical substances have legitimate use. Arsenic and strychnine in small doses were once used for medical purposes. Some of the substances are insecticides and others are compounds employed as cleaning agents, herbicides or rodenticides. If terrorists want they can manufacture potent chemical poison from such substances, which are readily available, or from various pesticides and herbicides. The technical knowledge and experience needed to produce a chemical weapon is less than that required for manufacturing a biological weapon and virtually all the materials and equipment required can be bought commercially.[23] In waging a chemical warfare, the real difficulty is not in manufacture or acquisition of poison but in dissemination. Chemical weapons are affected by the direction of the wind as well as temperature of the air. Dispersion by means of aerosol always involves a high loss of toxicity, which leads to reduction in lethality of the agent. Large number of actual attempts to use chemical weapons have been made by terrorists in various countries, but with less number of casualities due to insufficient amounts of chemical agents being used and most of the attempts were amateurish. However, the fear of poison is great and panic caused by an attack of this kind is enormous. It might cause more psychological than physical havoc.

A majority of nations are parties to the Chemical Weapons Convention (CWC), which prohibits the development, production, acquisition, retention, stockpiling, transfer and use of chemical weapons; but it is unlikely that this treaty has ended potential chemical weapons threat to the world community. In 1999, the intelligence community assessed that, despite the CWC and related supplier restraint regimes such as Australia Group, at least 16 states maintained active, clandestine chemical weapons programme.[24] The challenge to non-proliferation regime becomes even more acute in combating chemical weapon development, since many chemical warfare agents are simple to produce. They are often based on technology that is at least 80 years old and in some cases even older, putting them well within the reach of any Third World country that wants them. Although, newer agents, such as the reputed Russian developed Novichok class of next generation nerve agents may not be as readily accessible, the "technology for the agents is widely available in public domain".[25]

**Table 3.3 Difference between Biological and Chemical Agents**

| Characteristics | Biological Agent | Chemical Agents |
|---|---|---|
| Toxicity | Small amounts can kill. It is measured in pico grams (1 x 10 raised to the power – 12 grams) | Measured in milligrams |
| Speed of Action | Slow–several hours to weeks. Anthrax itself will take one to six days. | Kills within minutes |
| Specificity | Affects human beings and plants | Affects all living and non-living things. |
| Controllability | Effects dependent on meteorological conditions, including wind. | Dependent on meteorological and terrain conditions. |
| Residual Effects | Limited survivability | Very limited survivability. |

## Toxic Industrial Chemicals

The ease or difficulty for terrorists to cause mass casualties with an improvised Chemical or Biological weapon depends on the agent. "Toxic Industrial Chemicals" (TICs) can cause mass casualties and require little, if any, expertise or sophisticated methods. Most scholars on terrorism consider chemical warfare (CW) agents to be the weapon of choice given their lethality. They overlook an important point. Purchasing or stealing toxic industrial chemicals and using them as the weapons of mass destruction are an easier alternative. For instance, CX or phosgene oxime, one of the original Chemical Warfare agents, is widely known as a TIC. It is manufactured, stored, and used throughout the country just like other TICs. Very toxic insecticides are essentially unregulated items and almost as toxic as CW agents.

Truckloads of TICs are on the highways daily and a hijacked truckload poses a potential threat.

Most nations do not have an effective defence against chemical terrorism because they assume that the enemy will use CW agents rather than toxic industrial chemical compounds. There is a need to realistically assess the threat and risk of chemicals that are most likely to be used. There is a need to assess and evaluate the risk of any chemical to function as a weapon of mass destruction.

There are several advantages of toxic chemicals to make their use attractive for terrorists who seek to cause mass casualties. These are:

(a) The effect or symptoms can be delayed for a period of time depending on the chemical.

(b) There is minimal risk of detection.

(c) The relative ease of manufacture or acquisition.

(d) Hazardous or extremely hazardous substances are used in a number of chemical industries where there may be a breach in or a persistent lack of security.

(e) These hazardous chemicals are transported daily in vehicles, which can be hijacked.

(f) Trucks / vehicles can be used by suicidal drivers to transport the weapons manufactured in the illegal factories to their target.

***Possible Targets***: Food is the quintessential target for any terrorist group that seeks to kill large numbers of citizens and to produce widespread panic on the survivors. The fact is that a single terrorist needs only to contaminate a small quantity of each produce or beverage to be successful. The goal is to convince the public that the entire food supply is contaminated, with the subsequent loss in confidence in the integrity of the food system. Terrorists in possession of certain lethal chemicals, such as sodium monofluoroacetate and 8-fluorooctanol, have the ability to poison a large number of civilians in a simultaneous attack. Sodium monofluoroacetate is a colourless, odourless, tasteless and water soluble poison which, in the hands of terrorists working in restaurants, supermarkets and places where food is processed, is by any definition a weapon of mass destruction. What makes it so terrifying and deadly is the fact that there is no antidote and victims die slowly. In addition, there are no symptoms until an hour or two after initial exposure. The terror starts with vomiting, hypertension of the limbs, convulsions, collapse and it ends within seven hours. A plausible scenario as an example to illustrate the ease with which a single chemo-terrorist could inflict widespread panic with a simple and popular beverage on any morning. A 10 Kilogram sack of ground coffee or tealeaves yields 800 cups of coffee. Injection of 35 ml of 8-fluorooctanol into the sack before delivery to the user results in one LD50 per cup. The coffee or tea appears normal, the brewing process does not destroy the weapon of mass destruction, and the presence of the poison is not detectable by taste, odour or appearance. Contamination of seven sacks is sufficient to kill 2,800.

***Future Use***: Toxic Industrial Chemicals (TICs) are weapons of mass destruction and are readily available to modern terrorists. Several religious fanatics in possession of lethal chemicals that are presently undetectable are capable of inflicting large numbers of casualties on a naïve and unsuspecting population unprepared for asymmetric chemical warfare. Anyone who is capable, with complete self-control, of turning his or her body into ash, is also capable of carrying a backpack containing a ternary weapon system comprised of powdered zinc, concentrated hydrochloric acid and arsenic trioxide to generate enough arsine gas to kill thousands; arsine is a colourless gas with a toxicity threshold of 0.5 pp/hr.

## Biotechnology and Bioterrorism

Biotechnology is a broad term used to describe the production of innovative products, devices and organisms by exploiting biological processes.[26] In modern biotechnology, genes and genetic structure of cells is manipulated to create organisms. Biotechnological capacity is increasing and spreading rapidly.

This trend seems unstoppable, since the economic, medical and food-security benefits of genetic manipulation appear so great. As a consequence, thresholds for the artificial enhancement or creation of dangerous pathogens – disease causing organisms – will steadily drop. The revolution in biotechnology will therefore almost inevitably place greater destructive power in the hands of smaller groups of the technically competent: those with skills sufficient to make use of advances of the international scientific community. This future is being driven not primarily by military programmes, but rather by open, legitimate private and academic research.[27]

Human beings have engaged in genetic engineering for millennia, breeding new strains of both plants and animals. However, the discovery of the structure of DNA in 1953 and almost twenty years later, the invention of recombinant DNA technology permitted for the first time the direct manipulation of the genomes of organisms to achieve specific goals.28 Biotechnological inventions, such as the polymerise chain reaction (PCR) in the 1980s, 'DNA shuffling' in the 1990s and other technologies that automate copying, mutating, sequencing and otherwise manipulating DNA, introduced enormous numerical power into molecular biology. The resulting techniques of directed molecular evolution permit investigators to evolve capabilities without having to know beforehand what exact DNA modifications are necessary to achieve the desired result. Instead, it is now possible to engineer via trial and error, in which each step can involve populations of up to trillions of trial DNA molecules. Simultaneously, automation of key techniques has substantially lowered the threshold for performing these experiments. In part as a consequence of these advances, between 1993 and 1999 the biotechnology industry has spread rapidly around the world. For example, China now has some 20,000 people working in 200 biotechnology laboratories.[29] In January 2004, a member of the Pakistani Atomic Energy Commission (PAEC) concluded a two-week training workshop on 'Advanced techniques in biotechnology' in Faisalabad with the statement that the PAEC is committed to training scientists from Muslim countries in biotechnology.[30]

The globalisation of biotechnology is being driven not only by the national decisions, but also by biotechnology firms' growing use of international subcontracting and technological cooperation agreements to further their research – including their biodefence-related research, such as vaccine development for potential bio-terrorist agents.

The Human Genome Project and Human Genome Diversity Project are carrying out research on genetic difference between groups. This will provide scientists the human genome blueprint or genetic code of different ethnic groups, thereby enabling specific targeting of a group, providing to the terrorist a flawless weapon for "ethnic cleansing".[31] Thus, the most dangerous use of genetically engineered attacks would be when these are targeted against an ethnic group by designing a bacteria or a virus which could affect only that group, say non-whites in America, the Serbs or even people with blue eyes.[32] Genomic or the science and technology that allows information about DNA sequencing of genes to be translated into an understanding of their function and behaviour will enable not only understanding of the functioning of the human body but also enable us to disturb it, which provides the potential for its use in biological warfare.

The British Medical Association conducted a study that brought out that weapons that can distinguish between ethnic groups by exploiting the genetic or cellular differences could be a reality within the next ten years (by 2010). This will be "precision guided terrorism" and would remove one of the main barriers to employment of biological weapons – the risk of infecting friendly forces and members of one's own sect. The Human Genome Project could provide scientists with the genetic imprint of different ethnic groups, giving them a clue on how to attack a specific group only. The panel convened by the British Medical Association held that viruses and other microorganisms tasked to detect differences in the DNA of races could offer new means for ethnic cleansing. Such developments have also been reported in South Africa where scientists of the South African Biological Weapons programme were asked to develop ethnic weapons in the dying years of Apartheid.[33]

It is important to distinguish between diseases that are contagious – capable of spreading from person to person, like smallpox, influenza or pneumonic plague – and those that are not (such as anthrax). To cause mass casualties, non-contagious biological agents will still require aerosolisation – the processing of the agent into an extremely fine powder or mist, and its effective dissemination. Aum Shinrikyo's failure with this technology is a reminder that, without specialised knowledge or a research and development programme, mastering these techniques is not easily done. However, contagious agents could be used by terrorists to cause widespread outbreaks on the basis of person-to-person transmission, possibly skipping the aerosolisation step. Given this fact, the kinds of experiments surveyed suggest that, in future, in the realm of biology, we could face a 'banalisation' of WMD capacity. Many laboratories may have the capability to modify or synthesise deadly contagious pathogens, albeit with substantial risk to the researchers in the absence of proper containment facilities. The increasing ease with which microorganisms may be given deadly modifications emphasizes how even more misleading the moniker 'weapons of mass destruction' (WMD) will become in the future.

The nuclear, biological and chemical agents constitute the weapons of mass destruction (WMD) and these weapons have long been a subject of human imagination. The use of biological and chemicals weapons in war can be traced back for centuries; nuclear weapons of course, appeared more recently. At the advent of twentyfirst century all these weapons, particularly biological/chemical are increasingly available to more states and, what is more frightening, to small groups and even individuals. It is important for all us to study the characteristics of these weapons so that mankind is better prepared firstly to prevent the weapons technologies falling into the hands of non-state actors and secondly to manage the consequences in case terrorists are able to use these weapons against mankind.

## References

1. ANF Mullins, *"Weapons of Mass Destruction in the Developing World: The Diffusion of Advanced Weaponry"*, in W Thomas Warder (ed), American Association for Advancement of Science (Washington, 1994).
2. Lawrence Freedom, *"The Third World War"*, Survival, Vol. 43, No. 4 (2001), p.73.
3. JCS J-3 *Weapons of Mass Destruction (WMD) Handbook,* Washington, DC (U.S. Department of Defense, February 2001), p.7.
4. Jessica Stern, *"The Ultimate Terrorists"* (Harvard University Press, London 1999), pp.60-68.
5. Alex P Schmidt, *"Terrorism and The Use of Weapons of Mass Destruction: From Whom the Risk"*, (London: Frank Cass2000), p.119.
6. Matin Zuberi, *"WMD in the Hands of Non-State Actors: An Emerging Threat`Assessm ent of the Possibility and its Impact on National Security,"* USI Journal, (July-September 2002), p.364.
7. "Walter Laqueur, *"The New Terrorism,"* (Phoenix Press, London 2001). p. 21.
8. Mohamed ElBaradei *"WMD and Global Security: Time for Reform"*, IAEA Bulletin, Volume 45, No.1 (June 2003), p.5.
9. Zuberi, n.6, p.362.
10. Fritz Steinhausler and George Bunn, *Protecting the Source,* IAEA Bulletin Vol.45, No.1 (June 2003). p. 18.
11. Lothar Koch *"Traces of Evidence: Nuclear Forensics and Illicit Trafficking"*, IAEA Bulletin Vol.45, No.1 (June 2003). p. 23.
12. Charles D Ferguson, *"Reducing the Threat of RDDs"*, IAEA Bulletin Vol.45, No.1 (June 2003). p. 13.
13. Gavin Cameroon, *"Nuclear Terrorism: A Threat Assessment for the 21st Century"* (Hampshire, London: Macmillan Press, 1999), pp 143-144.
14. Encyclopedia Britannica Inc, 1994-2001, CD ROM Edition.
15. Laqueur, n.7, p. 21.

16. United States Office of the Secretary of Defense, *"Reports of the Secretary of Defense's Adhoc Committee on Biological Warfare"*, 11 July 1949 (Declassified 25 June 1986), p.12 as quoted in Susan Wright (ed),*'Preventing a Biological Arms Race'*, (Massachusetts; Massachusetts Institute of Tecnology, 1990), p 175.
17. London Statement, p.1-3; US Government Printing Office, *"The Worldwide Biological Warfare Weapons Threat"*, (Washington, DC: 2001), p.1.
18. *Economic Times*, 2 November 2001.p.20.
19. United Nations General Assembly, Report of the Secretary General on Chemical and Biological Weapons and the Effects of Possible Use, A/7575 (New York; United Nations, 01 July 1969), p.6.
20. "Chemical Attacks", *Economic Times*, 29 October 2001, p.16.
21. Patterns of Global Terrorism-2000. US Department of State, April 2001 quoted by Andrew J Bacevich in *"Terrorizing the Truth"*, Foreign Policy (July-August 2001), p.75.
22. Gavin Cameroon, *"Nuclear Terrorism: A Threat Assessment for the 21st Century"* (Hampshire, London: McMillan Press, 1999), p.144.
23. Ibid, p.148.
24. John A. Lander, *"Unclassified Statement for the Record by Special Assistant to the Director of Central Intelligence for Non proliferation John A. Lander to the Commission to Assess the Organisation of the Federal Government to Combat the Proliferation of Weapons of Mass Destruction"*, 29April, 1999, pp.1 and 3.
25. US Government Printing Office, The Biological & Chemical Warfare Threat, rev ed (Washington, DC. 1999), p.32; Office of the Secretary of Defence, Proliferation Threat and Response, p. 4.
26. Malcolm Dando, Biological Warfare on 21st Century (London: Brasseys, 1994), p. 100.
27. Christopher F Chyba and Alex L Greninger, *"Biotechnology and Bioterrorism: An Unprecented world"*, Survival, Vol 46, No.2. (Summer 2004), pp. 143 – 162.

28. Christopher F Chyba, *"The Recombinant DNA Debate and the Precedent of Leo Szilard"*, in Sanford A Lakoff (ed), Science and Ethical Responsibility (New York; Addison - Wesley, 1980), pp. 251 – 264.
29. David Barboza, New York Times, 21 February 2003.
30. Staff report, *The Daily Times*, Lahore, 17 February 2004.
31. Schmidt, n 5, p.121.
32. Wendy Burnaby, The Plague Makers, (London: Vision Paperback, 1999), p.136.
33. Schmidt, n 5, pp.120-122.

# 4

# National Strategy to Combat WMD

*We shall require a substantially new manner of thinking if mankind has to survive.*

**Albert Einstein**

The enemy today is terrorism–premeditated, politically motivated violence perpetrated against non-combatant targets by sub-national groups or clandestine agents. Those who employ terrorism, regardless of their specific secular or religious objectives, strive to subvert the rule of law and effect change through violence and fear. Terrorists also share the misguided belief that killing, kidnapping, exhorting, robbing and wreaking havoc to terrorise people are legitimate forms of political action. The likelihood of possession of Weapons of Mass Destruction (WMD)–nuclear, biological and chemical by terrorists has further aggravated the threat of terrorism and mass destruction.

The struggle against terrorism is different from any other war in our history. Triumph over this cannot be achieved solely or even primarily through military might. Terrorist networks and their supporters will have to be fought by using every instrument of national power – diplomatic, economic, law enforcement, financial, information, intelligence and military. Progress will come through the persistent accumulation of successes – some seen, some unseen. The need is to always remain vigilant against new terrorist threats. An effective strategy for combating WMD terrorism should therefore form integral component of our

National Security Strategy. The strategy calls for direct and continuous action against terrorists groups, the cumulative effect of which will initially disrupt, over time degrade and ultimately destroy terrorist organisations. The more frequently and relentlessly one strikes at terrorism across all fronts, using all the tools of statecraft, the more effective it will be. To succeed, we must take full advantage of today's opportunities including the application of new technologies, increased emphasis on intelligence collection and analysis and strengthening of new partnerships and relationship with other countries. WMD in the hands of terrorists could enable adversaries to inflict massive harm on India.

## Nature of Terrorist Threat

Terrorists have left their mark in some way upon every country in the world. The worst single terrorist incident in terms of casualties has been the terrorist attacks on WTC Towers in New York City on 11 September 2001 in the United States of America in which 3038 lives of citizens from some 90 countries were lost. This incident was followed by the 2004 terrorist attack in Beslan in Russia, which claimed 338 lives, mostly of children, being the second worst terrorist incident. Though the number of terrorist attacks globally has been on the decline, the number of casualties annually has been on the rise. While terrorist attacks claimed an average of 258 lives every year for the 30-year period from 1968 to 1997, the figures from 1998 to 2003 shoots up to 2,174.[1]

Other than human casualties and those injured, terrorist's attacks can cause enormous financial damage too. The most financially damaging attack by far was the September 11 WTC attack which caused a financial loss of $20 billion of insured property damage and an estimated $ 100 billion of economic loss over time.[2]

## The Structure of Terror

Despite their diversity in motive, sophistication and strength, terrorist organisations share a basic structure as depicted in Figure 4.1.

**Figure 4.1. The Structure of Terror**

At the base, underlying conditions such as poverty, corruption, religious conflict and ethnic strife create opportunities for terrorists to exploit. Some of these conditions are real and some perceived. Terrorists use these conditions to justify their actions and expand their support. The international environment defines boundaries within which terrorist's strategies take shape. As a result of globalization, the environment unwittingly provides access to safe havens, capabilities and other support to terrorists. The physical base to terrorists is provided unwittingly by states through ignorance, inability or intent both physical (e.g. safe houses, training grounds and camps) and virtual (e.g. reliable communications and financial networks) – that terrorists need to plan, organise, train and conduct their operations. Once entrenched in a safe operating environment, the terrorist organisation begins to consolidate and expand. The terrorist organisation's structure, membership, resources and security determine its capability and reach.

At the top of the structure, the terrorist leadership provides the overall direction and strategy that links all these factors and thereby breathe life into terror campaign. The leadership becomes catalyst for terrorist action. The loss of leadership can cause many organisations to collapse. Some groups however, are more resilient and can promote new leadership should the original fail or fall. Still others have adopted a more decentralised organisation with largely autonomous cells, making our challenge even greater.

## The New Global Environment

With the end of Cold War, we have seen dramatic improvement in the ease of transnational communications, commerce and travel. Unfortunately, the terrorists adapted to this new international environment and turned the advances of the 20th century into the destructive enablers of the 21st century. Today, the terrorists coordinate their global activities through the use of personal couriers and communication technologies emblematic of our era-cellular and satellite phones, encrypted e-mail, Internet chat rooms, videotape and CD-ROMs. The terrorists have also exploited the international media to project their image and message worldwide. Terrorists pay their way with funds raised through front businesses, drug trafficking, credit card fraud, extortion and money from covert operations. Some of the terrorists groups are known to use ostensibly charitable organisations and non-governmental organisations (NGO's) for funding and recruitment. Money for their operations is transferred surreptitiously through numerous banks, money exchanges and alternate remittance systems (often known as "Hawalas") – some legitimate and unwitting, others not.[3] Terrorists can now use the advantage of technology to disperse leadership, training and logistics not just regionally but globally. Establishing and moving cells in virtually any country is relatively easy in a world where more than 140 million people live outside of their country of origin and millions of people cross international borders every day.

Furthermore, terrorist groups have become increasingly self-sufficient by exploiting the global environment to support their operations. Whether it is the al-Qaeda's profiting from the poppy fields in Afghanistan, or terrorists operating in Indian State of Jammu and Kashmir and Northeast being funded by ISI which gets its funds from drug trafficking or Abu Sayyaf's kidnapping for profit in the Philippines, terrorists are increasingly using criminal activities to support and fund terror. In addition to finding sanctuary within the boundaries of a state sponsor, terrorists often seek out states where they can operate with impunity, because the central government is unable to stop them. More audaciously, foreign terrorists also establish cells in the very open, liberal, and tolerant societies they plan to attack.

## Availability of WMD and India's Security

Weapons of mass destruction in the possession of hostile states in India's neighborhood, their tendency to proliferate WMD and terrorists represents one of the gravest security challenges facing India and the entire international community today. The probability of a terrorist organization using a chemical, biological, radiological, or nuclear weapon, has increased significantly during the past decade. The availability of critical technologies, the willingness of some scientists and others to cooperate with terrorists and the ease of intercontinental transportation enable terrorist organisations to more easily acquire, manufacture, deploy and initiate a WMD attack against a nation state. Motivated by extreme, even apocalyptic ideologies, some terrorists' ambitions to inflict mayhem seems unlimited. The Aum Shinrikyo's unsuccessful efforts to deploy biological weapons and its lethal 1995 sarin gas attack in Tokyo subway provided an early warning of such willingness to acquire and use WMD.[4] In 1998, Osama bin Laden proclaimed the acquisition of WMD a "religious duty" and evidence collected in Afghanistan proves al-Qaeda sought to fulfill this "duty".[5] The threat of terrorists acquiring and using WMD is a clear and present danger. A central goal must be to prevent terrorists from acquiring or manufacturing the WMD that would enable them to act on their worst ambitions.

**Strategic Intent:** The intent of our national strategy should be to stop terrorist attacks against the citizens and our interests. To accomplish this task, there is a requirement to act simultaneously on four fronts.

The terrorists must be denied sponsorship, support and sanctuary. The UN Security Council Resolution 1373 and the 12 UN counter terrorism conventions and protocols establish high standards. In concert with other countries, especially with neighbours, a state should be able to coordinate its actions and ensure that they are mutually reinforcing and cumulative. Strategy's second front stresses denying terrorists the sponsorship, support and sanctuary that enables them to exist, gain strength, train, plan, and execute their attacks. The strategy to deny sponsorship, support and sanctuary focuses on the responsibilities of all states to fulfill their obligations to combat terrorism within their borders. The goal of this aspect of National Strategy is to choke off the lifeline of terrorists groups - their access to territory, funds, equipment, training and unimpeded transit. In the new global environment it is important for the state to understand how terrorists and their supporters may use legitimate means of communication, commerce and transportation for illegal activities. The UN Security Council Resolution 1373 clearly establishess the states' obligations for combating terrorism. The resolution calls upon the all member states to cooperate to prevent terrorists attacks through a spectrum of activities including suppressing and freezing terrorists finances, prohibiting their nationals from financially supporting terrorists, denying safe haven, and taking steps to prevent movement of terrorists. Additionally, the 12 international counter-terrorism conventions and protocols, together with UN Security Council Resolution 1373, set forth a compelling body of international obligations relating to counter-terrorism.

The state must be prepared to interdict terrorists ground, air, maritime and cyber traffic by positioning forces and assets to deny terrorists' access to new recruits, financing, equipment, arms and information. This effort will be specifically required to combat WMD threat by executing timely, effective interdiction effort against WMD related materials, technologies and expertise.

Another aspect of this strategy calls for preventing the laundering of drug trafficking finances through international financial system. The nation-state should evolve plans to prevent transit and sale of drugs through its territory by having effective intelligence and a national drug control strategy.

Some weak states /countries, though committed to fight terrorism, lack the capacity to fulfill their sovereign responsibilities due to lack of military, technical and training capabilities. They should be helped to acquire the necessary capabilities to fight terrorism through a variety of means including technical assistance, intelligence sharing, weapons and military training. An example of this cooperation is the active help being provided by India to Nepal to deal with the serious Maoist problem being faced by the latter. India views the insurgency in Nepal as a "common security challenge" since there are strong linkages between the Maoists and the left-wing extremist groups in India like People's War and MCC.

The second element of the strategy calls for a need to diminish underlying conditions such as poverty, hunger and deprivation of the masses that terrorists seek to exploit by focusing on the development and financial needs of the areas most at risk. These can be achieved by providing adequate aid to the region by fostering economic, social and political development, good governance and rule of the law. Aim being to address the underlying conditions that terrorists often seeks to manipulate for their own advantage.

The efforts to diminish the underlying conditions of people living with poverty, deprivation, religion and unresolved political disputes including perceived discrimination against them, due to their location have material as well as intangible dimensions. The nation must make efforts to resolve the regional disputes, foster economic, social and political development, religious freedom, good governance and the rule of the law, while not necessarily focused on combating terrorism, contribute to the campaign by addressing underlying conditions that terrorists often seek to manipulate to their own advantage. Every state should be capable of looking after its own people - their welfare, health, prosperity and freedom - and control its borders.

The government should work in a sincere manner to solve the perceived grievances of a particular segment of its population.

The underlying problems, which fan various terrorists' movements, should be addressed systematically in a timebound programme. Necessary legislation should be passed, which should be binding on all for the proper development of the regions that feel alienated from the mainstream. A change of government / ruling party should not effect the roadmap which has been drawn for the development of the region/population. There should be a comprehensive border management strategy, which will greatly enhance the ability of the state to screen, verify and process the entry of the people and goods into the country. The media plays an effective role to promote free flow of information and ideas to kindle hopes and aspiration among the population and it must be involved in reporting the good work being done by the state for the well being of its people.

The third element of the strategy calls for defeating the terrorist organisations by attacking their sanctuaries, leadership, command, control and communications, financial and material support. This approach will have a cascading effect across the larger terrorist landscape by disrupting their ability to plan and operate. An analysis of the history of combating terrorism confirms that the best way to defeat terrorism is to isolate and localise its activities and then destroy it through intensive, sustained action. Defeating terrorist organisations of global reach calls for direct or indirect use of diplomatic, economic, information, law enforcement, military, financial, intelligence and other instruments of power. However, due to the broad expanse and sophistication of some of the terrorists' organisations, we must first act to reduce their scope and capability. This calls for identification of the terrorists, locating their sanctuaries and destroying their ability to plan and operate. Preventing terrorist groups from gaining access to technology, particularly that which supports WMD should be accorded highest priority.

"Know your enemy" is one of the most accepted maxims in warfare. For the purpose of identifying, locating and then destroying the terrorists and terrorist organisations, the intelligence and the law enforcement agencies have a major role to play.

The information required would have to be gathered by exploiting the technological edge that we have over the terrorists' organisation. However, ultimate success can only be achieved by obtaining intelligence through human beings i.e. penetration of the organisation by human sources to provide information on the terrorists leadership plans, intentions, modus operandi, finances, communications and recruitment. Once the terrorist and his organisation has been identified and located, all resources available to disrupt, dismantle and destroy their capacity to conduct acts of terror must be utilised.

The fourth element of the strategy is to defend our own citizens and interests by both proactively protecting our state and extending our defences to ensure that we identify and neutralise the threat as early as possible. Through sustained effort to compress the scope and capability of terrorists' organisations, one can achieve victory against terrorists.

The final tenet of the strategy encompasses our nation's collective efforts to defend the country's sovereignty, territory and its national interests. The tenet includes the physical protection of the state, its populace, and property, as well as the protection of its interests.

We face an adaptive enemy. Empowered by modern technology and emboldened by success, terrorists seek to dictate the timing of their actions while avoiding our strengths and exploiting our vulnerabilities. Embodied in this strategy is the old adage that the best defence is a good offence. By improving and coordinating our indications and threat warnings, we will be able to detect terrorist plans before they mature. Through the continuous law enforcement, intelligence and military pursuit of terrorist organisations, the state can disrupt their ability to execute attacks. Our response to this complex mission requires a coordinated and focused effort from the whole nation, the central, state and district level administrations, the private sector and the people.

The capabilities of our agencies tasked with the protection of the population needs to be improved. Of interest will be vast improvement that would be required for collection, analysis and dissemination of information/intelligence, amongst all as quickly as possible. The procedures and systems that facilitate inter-agency, inter-governmental and private information sharing require to be expanded to all agencies to have access and input as necessary.

## Pillars of National Strategy to Combat WMD Terrorism

After having analysed the Strategic Intents of national strategy, there is a need to identify the pillars to base our strategy to combat WMD terrorism threat to India. These are highlighted in the subsequent paragraphs.

***Improved Intelligence Collection and Analysis***: An accurate and complete understanding of the full range of WMD threat is, and should remain among our highest intelligence priorities, to enable us to deter and defend against those who would use WMD. Particular emphasis must be accorded to improving intelligence regarding WMD related facilities and activities, interaction amongst our intelligence, law enforcement and military agencies and intelligence cooperation with friendly countries.

***Preparation:*** The possession and increased likelihood of use of WMD by hostile states and terrorists are realities of the contemporary security environment. It is, therefore, critical that armed forces and the civilian agencies are prepared to deter and defend against the full range of possible WMD employment scenarios. There is an urgent need to ensure that all the capabilities needed to combat WMD are fully integrated into our plans. This needs to be also fully integrated into basic doctrine, training and equipment of all forces so as to ensure that they can sustain operations to decisively defeat WMD-armed adversary.

***Interdiction:*** In collaboration with other nations we should join in effectively interdicting the movement of WMD materials, technology and expertise. Enhancing the capabilities of our intelligence agencies, armed forces and law enforcement agencies would greatly help them to effectively carry this out.

***Deterrence***: Today's threats are far more diverse and are far less predictable than those of the past. A strong declaratory policy and effective military forces are essential elements of our contemporary deterrent posture. An effective deterrent posture against WMD threats is reinforced by effective intelligence, surveillance, interdiction and domestic law enforcement capabilities.

***Defence and Mitigation***: Because deterrence may not succeed and due to potentially devastating consequence of WMD use against our forces and civilian population, Indian forces should have the capability to defend against such a threat including initiating pre-emptive measures. This calls for requisite capabilities to detect and destroy the WMD before they are used. In addition, robust active and passive defences and mitigation measures must be put in place to enable our forces to accomplish their mission. An effective response requires rapid attribution and robust strike capability.

***Research and Development***: There is a critical need for cutting edge technologies that can quickly and effectively detect, analyse, facilitate interdiction of, defend against, defeat and mitigate the consequences of WMD. If need be, we should take help of friendly foreign countries to possess these capabilities.

***Consequence Management***: India must be prepared to respond to the use of WMD against her citizens and armed forces. We need to develop and maintain the capability to reduce to the extent possible the potential horrific consequences of a WMD attack on our soil. There would be a need to conduct various programmes for planning, training and assistance to the state governments and district administration. To maximise their effectiveness, these efforts need to be comprehensive, well integrated and also well coordinated.

***Strengthened International Cooperation:*** WMD represents threat not just to India, but also as a whole to international community. For this reason, it is vital that we should work closely with likeminded countries on all elements to combat WMD menace. India should join the international community in its efforts to prevent terrorists from acquiring WMD. At the international level traditional measures such as diplomacy, arms control, multilateral agreements and export controls should be strictly regulated to dissuade or impede terrorist networks as well as to slow and make more restricted their access to sensitive technologies, material and expertise.

***Ensure An Integrated Incident Management Capability:*** The prevention of catastrophic terrorism is dependent upon interdiction of people and materials. However, solid plans, preparations and immediate response remain key to mitigating acts of terrorism.

Unity of effort requires coordination not only at the apex of the central government, but also at the operational/tactical level, where response and intervention actions may be taken by diverse authorities, acting independently or in coordination with each other. An effective, integrated response requires incident management planning, enhanced interoperability and coordination based on and supported by rapid and effective decision-making. In an effort to ensure rapid crisis response, there is need to develop plans for alerting, containing and if necessary, repelling an attack in progress while ensuring adequate resources are available to mitigate the damage by resorting to a well planned and rehearsed consequence management plan.

One of the most difficult challenges we face is to prevent, deter and defend against the acquisition and use of WMD by terrorist groups. The current and potential future linkages between terrorist groups and state sponsors of terrorism are particularly dangerous and require priority attention. The requirements to prevent, deter, defend against and respond to today's WMD threats are complex and challenging. The pillars of our National Strategy to combat the threat of WMD terrorism are seamless elements of a comprehensive approach. And they need to be very well integrated.

## References

1. Samiran Chakrawerttti, Death Toll in Terror Attacks on the Rise, *Times of India*, 07 Sep 2004. p.11.
2. Ibid, p.11.
3. Brigitte L. Nacos, *Mass Mediated Terrorism* (Rowman &Littlefield: Oxford, 2002), p.109.
4. Jessica Stern, *The Ultimate Terrorists*, (Harvard University Press: 1999), p.68.
5. Nacos, n.3, p.41.
6. Brigadier AK Lal, *Neo-Terrorism*, (Gyan Publishing House: New Delhi), p.207.

# 5

# Strategy for Consequence Management

*There is no single defence against this threat. Instead we must treat it as if it were a chronic disease, being constantly alert to the early symptoms and ready to employ, rapidly, a combination of treatments.*

**William S. Cohen, 1997**

The much-vaunted globalisation, new liberalisation, increasing privatisation of state functions and most importantly, the new revolution in computing, telecommunication and data transference capacities have all impacted strongly on the international security agenda and on the nature of the threat actors in today's world. The nation states today face asymmetric threats from non-state actors and terrorists. These include chemical, biological and radiological capable actors, aided by an ever-increasing number of extremists groups. Modern societies are particularly susceptible to weapons that are capable of killing many people at one time, as the citizens tend to live, work and travel in close proximity, thus providing them easy concentrated targets. Also, despite this concentrationof population, many individuals in modern societies are isolated from one another, thus creating a fertile ground for the breeding of extremists and making it possible for extremists group to operate unnoticed. As the constraints against terrorists using weapons of mass destruction seem to be gradually eroding, WMD are becoming valuable to terrorists who are religiously motivated and are seeking to conjure a sense of divine retribution, to display scientific prowess, to kill large numbers of people to invoke dread. The number of terrorists and terrorists groups motivated by objectives like these rather than the traditional objectives are increasing. Coupled with easy availability due to proliferation, advances in technology have made terrorism with weapons of mass destruction that are easier to use.

Despite these developments, terrorists' use of weapons of mass destruction is likely to remain rare. Few terrorists will be capable of using these weapons except in small-scale incidents, and few will want to kill tens or hundreds of thousands of people. It is useful to think of the danger in terms of the concept of expected cost: the product of the probability of an event and its consequences. While the probability of WMD terrorism is low, its expected cost - in lives lost and in threats to civil liberties - is potentially devastating. But the constraints against terrorist using weapons of mass destruction seem to be eroding gradually and therefore, there is a need to pursue a wide variety of strategies for preventing WMD attacks and for minimizing loss of life, if the attacks do occur.

Terrorism, by nature, is unpredictable. This makes the preparation for as well as response to incidents difficult. A terrorist incident is divided into two stages of management – crisis management and consequence management. While the distinction does not directly affect local emergency responders, understanding the response frameworks helps minimise confusion and potential role conflicts.

## Crisis Management

Crisis management entails those response measures or actions that are required to identify, acquire and plan the use of resources needed to anticipate, prevent or resolve a threat or act of terrorism; it is primarily a law enforcement response which focuses on criminal aspects of the incident. Crisis response refers to instances where the perpetrators of an assault have been discovered before an actual release; in this context, there could be one or more "devices" to be rendered safe as a result of uncovering the plan or neutralizing the terrorists. The primary authority/responsibility in these cases, will be that of the central government, which will take measures to anticipate, prevent and/or resolve a terrorist threat.

## Consequence Management

Consequence management involves measures to alleviate the damage, loss, hardship or suffering caused by emergencies. It includes measures to restore essential government services, protect public health and safety and provide emergency relief to individuals and business affected by deliberate or inadvertent consequences of a chemical, biological, radiological, nuclear or high-yield explosives event. Consequence management is basically preparedness for dealing with and responding to the consequences of an incident involving WMD; it describes ways or means to alleviate the short and long term physical, socio-economic, and psychological effects of a chemical, biological or radiological attack. It describes the coordination of international, national, regional, and local assets to deal with the effects of such an attack. The term includes preparatory work in response to a WMD threat that would include site surveys: assessments of the ability of local hospitals to treat or decontaminate victims, and the size, condition, and locations of local stocks of various antidotes. Preparation could include determining the locations, size, and availability of other national antidote stocks as well as international stocks available to support planning for surge capacity.

**Figure 5.1 Crisis Vs Consequence Management**

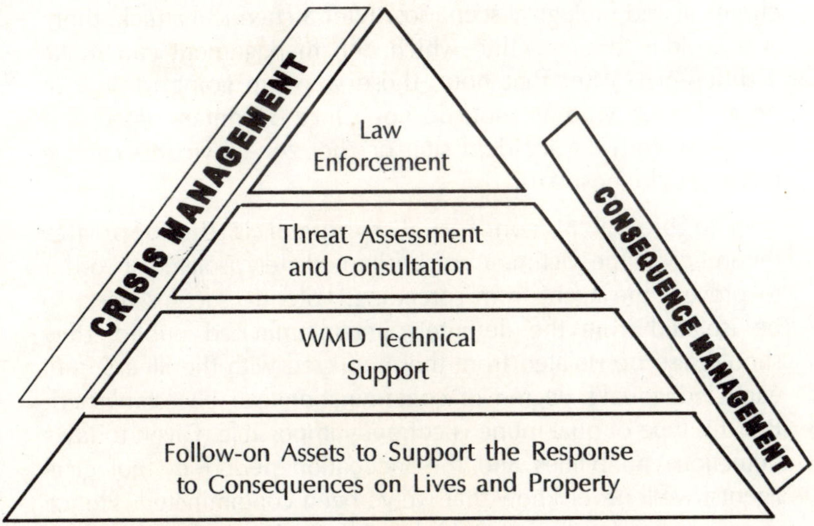


Source: United States office of Antiterrorism Assistance

Thereafter, should police agencies be unable to prevent the incident, consequence management would include treatment of victims within a contaminated zone, their decontamination and evacuation, and local cleanup. It would also involve psychological treatment and other efforts to restore confidence in the social and economic well-being of the affected area and of the country itself. The experience gained from past management of natural disasters including the Gujarat earthquake suggests that the latter would require long-term commitment of considerable resources.

From the foregoing, it is clear that crisis response and consequence management are parallel and overlapping and it should be clear that any division between response before the fact and dealing with consequences after the fact of an attack is both arbitrary and potentially misleading. It should also be apparent that actions taken in response to the consequences of a WMD attack would invariably have two kinds of outcomes. Not only will such actions have their own consequences, intended or otherwise - and which may or may not play into the hands of terrorist who launched the WMD assault- yet consequence management clearly cannot be conducted on an ad hoc basis. Consider for example, the contrast between chemical and biological scenarios. After a chemical attack, there is a "golden hour" within which deft management can make a difference. After that hour, those who are going to survive do and those who are not, do not. Once decontaminated and removed from the incident site, or "hot zone," victims can be dispersed to hospitals.

The biological conceptual approach is the opposite: theoretically, the victims must be immediately isolated in order to prevent the agent from spreading. Potential victims need to be isolated from the definitely uncontaminated public; they should also be isolated from those afflicted with the illness until each individual's degree of contamination can be established. But this type of quarantine is currently impossible. Given today's detection capabilities and the incubation period of biological agents, we'll never know that we've been contaminated. Hence the diabolical genius of a biological attack: we become "the (unknowing) vector of our own death."[1] Given the varied dimensions and manifestations of chemical and biological terrorism, the battle of consequence management would be won, if there are consultations and planning well before any threat of an incident emerges. We must, therefore, think of crisis response and consequence management as parallel and overlapping continuums that all agencies tasked to deal with WMD terrorism must keep constantly in view. In a WMD situation, consequence management is the crisis and is also an almost indescribably complex task.

## Concept Of Consequence Management

The response to an incident involving WMD is vertically organized from local to national level. The primary authority / responsibility is that of the state /local government, though whatever assistance is required by the state government is provided by the central government.

On occurrence of a WMD disaster/terrorist incident, the first

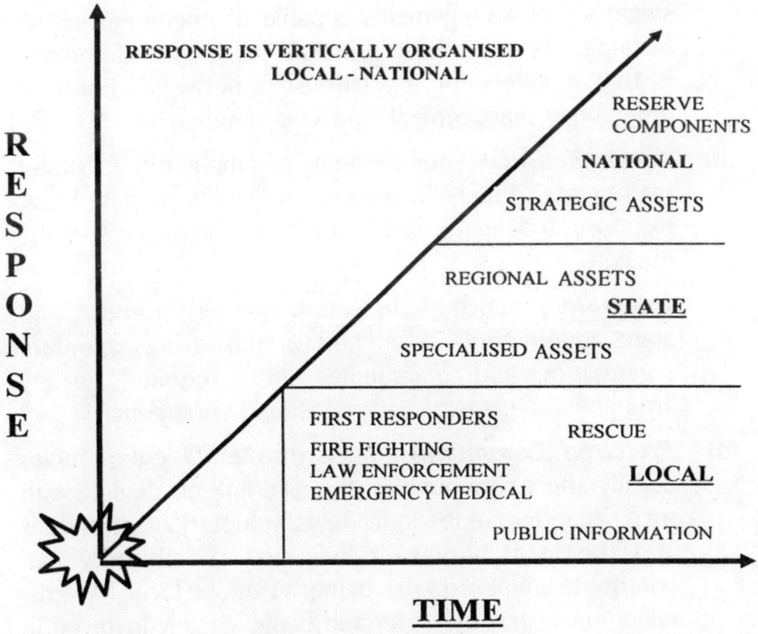


Graph 5.1 WMD Responses

to help the victims would be the local population of the affected area. With passage of information the First Responders will include local police along with the fire services; and the local medical authorities. These responders would be required to carry out various tasks, which have been explained in detail in Chapter on First Responders. Keeping in view the magnitude of the disaster, resources from the neighbouring districts/ states would respond by building up on the First Responders. At the state level, there would be specialised assets, which would respond with specialised manpower and equipment. At the national level, the Armed Forces of the country would respond with strategic assets. However, if the state /national assets are located in the near vicinity of the disaster site, they would respond immediately.

## Approach to Consequence Management

At various levels, our approach to Consequence Management should be four dimensional as under: -

(a) ***'All in One'***: Though specific counter-measures will often vary with different hazards, it is desirable to establish a single set of arrangements capable of encompassing all hazards. The underlying principle for such an approach is that it caters for a relationship between disaster / emergency management and civil defence.

(b) ***Comprehensive***: Four elements of emergency / disaster management, timely prevention / mitigation, preparedness, response and recovery are crucial to the success of the mission.

(c) ***Integrated:*** Active participation amongst Central and State governments, local bodies, industries, voluntary organisations and communities will be required, for any arrangements to deal with disasters / emergencies.

(d) ***Prepared Community***: Since disasters / emergencies usually affect communities, the vital link on dealing with such situations are the individuals, voluntary organizations and the local bodies. Individuals / community can contribute immensely by being aware of local hazards, taking protective measures and being actively involved in voluntary organizations as well as ensuring that local bodies have effective measures in place.

## Imperatives for Consequence Management

***Sound Intelligence Base:*** An accurate and complete understanding of full range of weapons of mass destruction threat is and should remain amongst the highest priorities, to be able to deter and defend against those who would use WMD. The particular emphasis will have to be laid on WMD-related facilities and activities. The sound intelligence databank could be facilitated by establishment of a 'National Terrorism Intelligence Centre' which would carry out the functions of networking of all domestic intelligence agencies, maintain a national level database and share the same, whenever required with International intelligence agencies.

The intelligence databank should provide liaison interface between various agencies (Industry-Scientific Community-Military-Government) and should have an integrated network for dissemination of information. It should also be able to provide realtime update of databank through domestic/external inputs and media interface for update and exchange of information.

***Pre-disaster Early Warning and Security Measures:*** Ensuring pre-disaster early warning and security measures will lead to saving of precious human lives. Security is an 'all embracing' concept and includes men, material and information. The elements of a security system are:

(a) ***Deterrence:*** Today's threats are far more diverse and less predictable. A strong declaratory policy and effective military forces are essential elements of an contemporary deterrent posture. An effective deterrent posture against WMD threats is reinforced by effective intelligence, surveillance, interdiction and domestic law enforcement capabilities. Deterrence can be ensured by denying access to secure areas by raising cost of violation with life and putting in place visible physical, electronic and human barriers activated for 'strike to kill'. Adequate strike forces equipped with sophisticated weapons should be catered for.

(b) ***Detection***: Layered detection sensors to include electric fences, buried seismic motion detectors, video surveillance / CCTV and alarm system would be adequate to protect the installations against terrorist attacks.

(c) ***Denial of Access***: The important facilities like nuclear power plants, chemical factories and biological research facilities should have restricted access. This can be ensured by having in place physical barriers through special passes/ badges/electronic card readers, on site security forces, off site security forces, quick reaction special response teams, two/three man rule to gain access and digital coded keys for selective access.

## Command and Control Set Up.

***National Level***: At the national level the central government should issue a National Emergency Management directive / plan to include:

(a) Provide a national framework for WMD terrorism prevention and mitigation.

(b) Review existing arrangements for preparedness.

(c) Inter-departmental interface.

(d) Lay down policies at national level.

(e) Delineation of responsibility to separate agencies for handling nuclear, biological and chemical terrorism incidents/disasters.

***Organisation Structure at National level***: The Ministry of Home Affairs (MHA) is responsible for disaster management in the country.

***Empowered Group of Ministers on Disaster Management (EGM):*** Headed by the Minister of Home Affairs, it comprises 14 additional members–Ministers from the Ministry of Rural Development, Railways, Coal and Mines, Power, Textiles, Consumer Affairs, Food and Public Distribution, External Affairs, Defence, Agriculture, Finance, Information & Broadcasting, Health & Family Welfare and the Deputy Chairman of the Planning Commission.

***National Crisis Management Committee (NCMC):*** A National Crisis Management Committee (NCMC) has been constituted in the Cabinet Secretariat under the Chairmanship of the Cabinet Secretary. The other members are Secretaries Home, External Affairs, Defence, Railways, Power, Petroleum & Natural Gas, Civil Aviation, Planning Commission, Secretaries of Department of Animal Husbandry and Dairying, Drinking Water and Supply Expenditure, Food and Public Distribution, Health, Road Transport and Highways and Rural Development and Headquarters Integrated Defence Staff. An officer of Cabinet Secretariat is the Convener. When a situation is to be handled also by the NCMC, it will give such directions to the Crisis Management Group of the Ministry as deemed necessary. The Director Disaster Management, MHA is responsible for prompt notifications of all ensuing development to the NCMC. Directions are given to ministries/ Departments/Organisations for specific action to meet a crisis situation.

***Crisis Management Group (CMG)***: A Relief Commissioner is appointed at the Centre, preferably a Special Secretary from MHA, to deal with the matters relating to relief in the wake of major natural calamities. The CMG consist of representatives of the level of Joint Secretaries in the Ministries/Department of Finance, Defence, Food, Civil Supplies, Railways, Power, Urban and Rural development, Health, Petroleum, Planning Commission, Tribal Affairs, Women & Child Development, ACIDS (Joint Operations), Animal Husbandry, Directors General of India Meteorological Dept and Civil Defence, and Communication and Surface Transport.

***National Emergency Management Authority (NEMA):*** At the national level, the day-to-day command, control and coordination of emergency response is overseen by NEMA. NEMA will take policy guidelines from the Crisis Management Committee, under Cabinet Secretary. All departments at the national level shall comply with instructions of NEMA. Headed by Special Secretary cum Chairman, NEMA Comprises of Special/ additional Secretaries from Health, Water Resources, Environment & Forests, Agriculture, Railways, Atomic Energy, Defence Research Development Organisation, Chemicals, Science

& Technology, Defence and Indian Meteorological Department. NEMA works under overall supervision of the Home Secretary. It has three divisions each headed by a Joint Secretary. These are:

(a) Disaster Prevention, Mitigation and Rehabilitation.

(b) Preparedness and Response.

(c) Human Resource Development administration, Parliament and Budget.

***Organisation at State and District Level:*** Presently the set up varies from State to State. There is a Special Committee under the Chief Minister for policy and monitoring. The Relief Commissioner/Revenue Secretary of the State is the nodal officer for directing relief effort. The DC/DM at District level is the focus point. However, NGO's, people's representatives and local bodies also get involved. The command, control and coordination of the emergency response will be with the State Dept of Disaster Management (SDDM)/ State Disaster Management Authority (SDMA) / State Emergency Management Authority (SEMA) and it shall function under policy guidelines by Chief Minister /Cabinet through the Chief Secretary.

## On Site Organisation

In case of a disaster/sabotage, involving industries dealing with hazardous radiological,chemical, biological and petrochemical materials, there would be no time for rescue and relief unless suitable teams are at hand. By the time National Emergency Teams fly into the area, reasonable amount of time would have elapsed and large-scale casualities may have already taken place, as in the case of Bhopal gas tragedy. As such, select personnel from affected industry should be organized as Territorial Army (TA) Units who should be trained and equipped to act as "First Responders" to undertake immediate deployment/actions at the place of incident/disaster. These will be primarily responsible for early warning, rescue, mitigation, neutralizing /capping the source if possible and coordinating the relief measures in conjunction with QRTs, local hospitals and other agencies.

The following Hazardous Industrial Units should have TA units:

(a) TA Unit (Radiological).

(b) TA Unit (Petrochemical).

(c) TA Unit (Chemical).

***Communication Requirements:*** Layered, redundant, survivable communications capable of realtime data transmission is of crucial importance. The requirement of connectivity will be, to have, vertical and lateral links for intelligence, law enforcement, fire fighting, on site security forces and for command and control set up for other agencies.

***Civil Defence Measures*:** Complete populations/assets cannot be protected, hence vulnerabilities need to be identified and prioritised. The towns could be categorized as under: -

(a) Vulnerability to external threat (proximity to border).

(b) Vulnerability due to proximity to nuclear installations.

(c) Vulnerability by way of population density /archeological /historical / political/ commercial importance.

(d) Vulnerability by way of seat of power/nodal centre for governance.

***Education of Population*:** In order to effectively prepare and deal with a WMD terrorism event, the education of the population with regard to the effects and methods of dealing with weapons of mass destruction plays a very important part. This will lead to prevention of panic and fear on occurrence of such an event. People could be educated by undertaking measures as under:

(a) Lectures in educational institutions.

(b) Adult awareness programmes.

(c) Awareness through media.

***Training*:** The requirement of training is complex, multi disciplinary and covers a very large domain. However, knowledge of unconventional weapons will help them to prepare for response to and recover from incidents involving such weapons. Meeting the threat from unconventional weapons requires police, fire department, medical and public health personnel to be trained. In addition, there is a need to train personnel from Non Government Organisations (NGO's) and the community. The training would be required in the following domains:-

(a) General education / awareness on WMD-related terrorism.

(b) Handling of hazardous chemicals.

(c) Equipment donning and operation.

(d) Safety precautions.

Today, we are living in an age of unprecedented changes where an individual or a small group of people possess the ability to wreak havoc on the population. This ability coupled with festering sores of tribalism, nationalism and elitism, forces us to protect ourselves from the effects of these new threats. With most of the constraints against terrorists using weapons of mass destruction eroding, there is a need for the State to pursue a wide variety of strategies for preventing WMD attacks and for minimizing loss of life, if attacks do occur. Having an effective consequence management strategy would act as a dependable deterrence.

# 6

# First Responders

*To send people into operations without training them is to abandon them.*

**Confucius**

The success of mitigating the consequences of a WMD strike depends upon the time taken by the First Responders to respond to the event. Time is a critical factor and rescue of victims must occur in the first hour, which is also known as the 'Golden Hour'. The rescue work will have to be carried out by the first responders without outside support and hence, there is need for these persons to be trained. Skills of responders are more important than the equipment required to manage the consequences of a WMD disaster/terrorist incident.

## WMD Response Foundation

Actions taken immediately after an emergency occurs most often have an impact the on outcome, success or failure of the operation conducted in response to a WMD incident, more than any other. The job of first responders is to save lives, protect property and preserve evidence-all steps that must be initiated by the first emergency service personnel who arrive on the scene. Every responder from the fire services chief down to the police officer or firefighter or emergency medical personnel should be thoroughly familiar with the actions required for effective management of a WMD incident.

An effective response foundation to prepare and deal with a WMD incident is laid on the training of the police, fire department personnel, hazard material technicians, emergency medical services, civil defence, public works, armed forces and other first-on-the-scene organisations. All those first responders should have the knowledge and a clear view of the response process.

**Figure 6.1 WMD Preparedness**

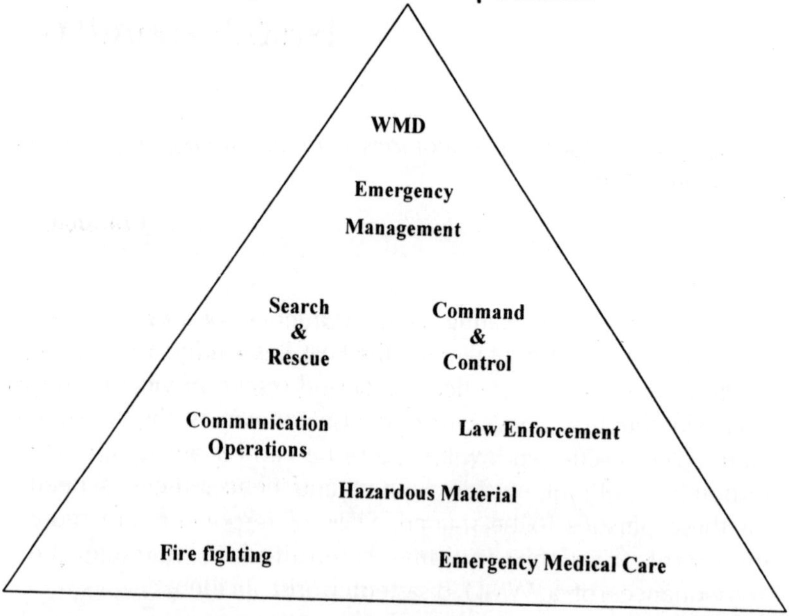


*Source:* United States Office of Antiterrorism Assistance Weapons of Mass Destruction (WMD) First Responder Training Programme.

For the response to be successful in a WMD incident, time is a critical factor. The rescue of victims in such a scenario must occur in the first hour. The first responders constituting of the police, firefighting and emergency medical care personnel will have to carry out the identification of the source of the hazardous material, carry out search and rescue operations, firefighting, provide first aid to the victims and inform the Disaster Management Authority of the district/ state. Communications would be required for the first responders to communicate amongst themselves and with outside agencies for requisitioning of aid and informing higher authorities. Laying down clear responsibilities and ensuring effective command and control at the disaster site will ensure an orderly emergency management. The actions would have to be carried out during preparatory stage. The police/law enforcement agency will also be required to preserve the evidence at the incident site to trace the culprits and bring them to book subsequently.

## First Responders

The first responders of any WMD-related disaster/incident are the affected local population. It is essential, therefore, that through regular awareness training and simulated exercises, the people are educated on the various actions to be taken in case there is such a happening. The media, particularly radio and television are ideal modes in building awareness among the common man. However, precaution should be exercised so as not to cause alarm.

***The Police Force:*** The first person to be contacted in case of any emergency is the policeman; therefore in case of a WMD disaster/incident also, he will be the first to be informed of any untoward happenings in his area of responsibility. The police can only react to such a scenario, if they are tasked, trained, and equipped to deal with such crisis. The police would be required to prepare suitable Standing Operating Procedures (SOP) for each police station listing out various actions to be taken. If required, necessary legislation to enhance Charter of Duties of police to deal with WMD crisis could be enacted. It should be clear to the police that WMD disasters/incidents like any other accidents are also their responsibility and their limits of responsibility should be unambiguously defined.

The police would be required to cordon off the affected area in two circles. The inner circle could be as close as possible to the area of the disaster and an outer cordon a few kilometers away to stop people from reaching the site of the disaster. This would also ensure that incoming routes are not cluttered with traffic, which would impede the movement or prevent assistance in the form of water tankers, ambulances and other forces required for disaster management. The second important task would be the setting up of an adhoc command and control structure. The Control Headquarters should be set up near a police check-post in a direction opposite to the direction of the wind.

***The Firefighting Organisation***: The firefighting personnel are the people who are trained and equipped to manage hazards. They would be required to carry out firefighting, rescue and evacuate marooned people to safe places. The fire department personnel would required to be in possession of detection equipment for identification of the type of chemical/ biological agents used in the incident by the terrorists. They would also required to be equipped with requisite clothing and other equipment to prevent them from becoming casualities themselves. An additional important task by the fire department would be to set up decontamination station to carry out decontamination of the people affected by various chemical/ biological/radiation agents.

***The Armed Forces***: The response of the military, particularly the army, will depend upon the proximity of the army units to the area of disaster. The whole country is divided into various army commands that are responsible for their nominated areas. Some cities and towns have military Station Headquarters while others, who do not have, are dependent on the military units/establishments located in their proximity. Civil–military liaison at state/district/city level for all purposes is an effective ongoing process. In addition, there are various Air Force and Naval Stations, which will provide necessary help in case of crisis. The armed forces also have well-trained and equipped medical teams. The armed forces are adequately trained and equipped to handle WMD disaster/terrorist strike. Further details are discussed in a separate Chapter.

***Medical Assistance***: On receiving intimation from police through the medical department of the town/city, the medical teams would rush to the site of disaster and report to the Control Headquarters set up by the police. The hospital staff would also be provided advance warning to be ready to receive the affected casualties. Once the teams deployed initially have been able to identify the type of agent used, the on-site Control

Room will report the same to the medical authorities and the hospitals so that adequate supply of antidote are ensured. Isolation wards must be vacated, reception and decontamination stations should be created and medical teams (less the medical personnel dispatched to the site) formed. Depending upon the magnitude of the incident/disaster hospitals in adjacent cities/ towns must also be warned and prepared to accept casualties.

***Civil Aministration:*** While the police, the fire department and the medical personnel are reporting in groups to the Control Headquarters, the civil administration should gradually take over control of the operations. The District Collector or the District Magistrate, as per the appointment, in that particular district will head the handling of the crisis. The district head will have to ensure smooth conduct of consequence management of the disaster.

***Civil Defence Organisation:*** The Civil Defence Organization of the district will have to be pressed into service at the earliest and will be mainly responsible for mitigation of the disaster by carrying out evacuation of casualties to places where immediate medical help is available and collecting the dead at earmarked place. Such people rendering relief will have to be properly equipped and protected against contamination. There is a need, therefore, for civil defence organizations to have NBC protective equipment.

***Non-Government Organisations (NGOs):*** NGOs have an important role to play in the management of WMD disaster/ terrorist strike situations. However, since they do not have the requisite training to operate in a contaminated environment, they will be best employed in tasks which are out of the contaminated area and away from contaminated personnel. They can best be employed on various other relief tasks.

## First Responder Requirements

One of the first and most important steps that emergency service personnel can take is to recognize that a terrorist attack has occurred. There are a multitude of indicators to warn the responder for each of the various types of weapons. Some of the details, which a first responder should know, are listed below:

(a) **Hazardous Material**

- Understand what hazardous materials are.
- Know the risks of hazardous materials.
- Understand outcome.
- Recognise a hazardous material release.
- Identify the hazardous material, if possible.
- Determine need for additional resources.
- Understand first responder roles.

(b) **Weapons of Mass Destruction**

- Understand WMD Agents.
- Understand risks of WMD agents.
- Recognise WMD release.
- Identify the WMD agent.
- Determine need for additional resources.
- Understand first responder roles.

## Difference Between Hazardous Material Incident and WMD Incident

Initially both the incidents look alike, however, a trained first responder like a police officer /firefighter will be able to distinguish between the two. The differences are:

(a) ***Hazardous Material Incident***: Generally, toxic material is involved and is easily identified by the responder.

(b) ***WMD Terrorism Incident***: It will be deliberate attack involving super toxic materials resulting in many casualties, which would require mass decontamination. It will be risky for the emergency first responders and civilians. In this case, the evidence preservation would be required at the crime scene and will involve major interaction / coordination between various agencies. The act is likely to result in chaos with the requirement of resources being overwhelmed. There is also a possibility of secondary device being activated, which would be designed to kill the first responders. There is also a likelihood of such incidents having pre-incident indication.

## Recognition of Devices

One of the first and most important steps that emergency service personnel can take is to recognise that a terrorist attack has occurred. There are a multitude of indicators to warn the responder for each of the various types of weapons. The most easily identified is the high yield conventional explosive device. In such an attack, it is most important to realise that the presence of a secondary device is highly feasible. Many a times this type of attack would require extensive immediate resources and result in diverse hazards, including collapse and fire. The other types of WMD attacks are much less apparent and require the responder to be acutely aware of small hints. Various recognition and other indicators specific to the type of device, are listed below.

## Conventional Devices

Conventional devices, are many times the most visually destructive and results in highest number of casualties on first observation. It must be remembered that the responder has a duty to look beyond this and to the needs of the incident. Upon arrival, the Incident Commander (IC) should thoroughly take in the whole situation and begin to implement the response functions discussed later in this chapter. Specific indications that a device of this type might have been employed include:

(a) Remains of incendiary devices.

(b) Odours of accelerants.

(c) Unusual heavy burning or fire volume.

(d) Large-scale damages to structures or vehicles.

(e) Blown-out windows.

(f) Widely scattered debris.

Medical indications include victims exhibiting symptoms of blast injuries including trauma from shrapnel and/or other bomb fragments, penetrating or blunt trauma from flying debris, a shock-like appearance, difficulty in hearing and damage to eardrums. Whenever the responder encounters such an incident, he/she should treat the location as a crime scene until evidence to the contrary is found. A secondary device should also be presumed present and to dispose off the device, steps should be implemented according to departmental procedures.

Standard conventional devices are most often considered, either improvised explosive devices (IED) or weapons grade explosive devices (e.g. Plastic explosives, dynamite). This definition must be expanded following the 11 September 2001 attacks in the United States to include other "non-conventional" means of attack that produce similar large-scale physical results. Although, the use of commercial planes have been employed in the past by terrorists to attack targets, the majority of the population limits their perceived threats to methods used in previous attacks. This no longer can be an assumption, especially for first responders.

Terrorists use methods of attack on targets to their advantage. The way in which they attempt to accomplish their goals will continue to evolve in order to catch us off-guard, persistently looking for the softest, or easiest places and ways to attack. The first responder should weigh the information available to them and determine through the indications listed above, whether to treat an attack such as on the 11 September WTC or Pentagon as a conventional device attack from an aircraft. In many respects, the overall indications, effects and functions of the first responder will remain the same.

***Nuclear Devices:*** Nuclear Devices, although rare and expensive to acquire, are a possibility in the current terrorists arsenal. The detonation of a nuclear device causes two hazards- the first is similar to that of a high yield conventional explosive (HYCE). Extensive damages will be present including fire and collapse hazards.

The difference between a nuclear and HYCE device is the presence of radiation. The effects of radiation can vary on humans depending on the level of their exposure to it. Most to effects are long term and do not present symptoms for an extended period of time following exposure. Specific instruments are needed to detect the type of radiation and determine the precautionary steps that should be taken to insure responder safety. Since these instruments are not readily available to first responders, it is imperative to request resources equipped to detect and handle a radiological incident at the first sign of their presence.

***Radiological Devices*:** Radiation itself is another form of WMD and can be dispersed without an explosion. Many times the presence of a radiological device will only be detected with specific instruments. With the difficulties in detecting this type of device, it is necessary to rely on the device being discovered and the quick reaction of responders to limit further exposure.

***Biological Devices*:** Another type of WMD that responders could come in contact with is a biological agent. There are two main presentations of biological agents. The first is through a community public health emergency. This presentation is when a device is employed and various numbers of citizens become infected or contaminated. As biological agents vary in incubation periods and symptoms, the period between infection and the start of a public health emergency can vary. The point where health officials determine the presence of an epidemic of biological infections, and their resources become strained, is the beginning of a public health emergency.

The second presentation is through a focused response to an incident. Many times, mdue to the difficulty and relative indiscrete nature of biological devices, the response to a biological incident may be from a specific threat or discovery of a delivery device.

In the instance of the September and October 2001 mailings of anthrax letters across the United States, in some of these cases no threat letter was enclosed. Here, the presence of a white powdery substance was the only indication of a biological agent. Only through education can such a case be treated as an emergency. However tragic these mailing and the successive cases of anthrax, it gained significant media coverage to educate the majority of population to be wary of such threats.

In either case, responders should approach the scene with both possibilities in their mind and prepare accordingly.

Whenever the presence of a biological agent is possible, an appropriate action to protect the responder and identify the agent is essential. Further, indications of biological agents include:

(a) Flu-like symptoms that may take days to weeks to show presentce.

(b) Unusual number of sicks or deaths.

(c) Dissemination of unscheduled and unusual sprays, especially outdoors and/or at night.

(d) Abandoned spray devices with no distinct odours or in unusual locations are also indications that should cause alarm and further inspection.

***Chemical Devices:*** Chemical incidents have their own distinct indicators such as rapid onset of similar symptoms in a large number of people, the presence of hazardous materials or lab equipment in unusual places, reports of unusual odours or tastes, explosions, mists, plumes or lingering gases. One indicator that may confuse the responder is an explosion that only destroys an individual package. With such an indication, the presence of a chemical or biological agent may be the primary threat and not the initial explosion.

As with biological agents, the unscheduled dissemination of unusual sprays or abandoned spray devices is also a warning sign. The main difference between chemical and biological agents is the time between delivery and the detection and alerting of emergency service personnel. In case of a chemical attack, large numbers of dead animals, mass casualties without apparent trauma or a distinct pattern of symptoms can be signs of a chemical attack versus a biological attack.

It is important to maintain familiarity with the indications for each type of WMD in order to quickly recognise its presence. The time it takes to determine the possibility of a WMD incident is of paramount concern. Time is of the essence – the longer it takes to make this determination, the larger the possible effect and longer the time for recovery.

## Threats to the Responder

Self-protection must be the primary concern of the responder. If the rescuers succumbs to the hazards present at an incident then they themselves become victims, further adding to the impact and consequences. For this reason, all emergency service personnel should not only be able to recognize the indications of a WMD, but also the effects on unprotected individuals.

***Threat Classification***: There are six main threats to which a first responder is likely to be exposed.

***Thermal***: Thermal harm can be inflicted through four main routes. The first, conduction, is the direct transfer of heat or electricity through a substance and can cause damages to skin and other contact membranes as well as internal organs. An example of conduction is touching the coil of an electric stove and suffering a thermal burn injury. Convection is the transfer of heat through the flow of liquid or gas. An example of a convection injury would be burns sustained from close proximity to a heat source.

***Radiation***: Radiation is the emission or transmission of energy in the form of waves through space or through a material medium causing spontaneous disintegration or decay of the nucleus of an atom. There are three levels of radiation with corresponding dangers and safety precautions: -

(a) ***Alpha***: Least penetrating and usually sustained by ingestion ossr inhalation. Use of protective clothing (PPE) reduces danger to Alpha radiation significantly.

(b) ***Beta:*** Higher wavelength of radiation, which has greater penetration and can cause skin or internal organ damage depending on proximity. PPE use can reduce danger to Beta radiation but does not provide long-term protection.

(c) ***Gamma***: Very short wavelength of radiation with significant penetrating power. Gamma radiation causes skin burns and can severely injure internal organs. PPE use does not provide safety from Gamma radiation.

***Asphyxiation:*** Asphyxiation is the ability of cells to receive oxygen. There are two types of asphyxiation: -

(a) **Simple:** Caused by the displacement of oxygen from haemoglobin, the human blood carrier by other inert gases.

(b) **Chemical:** Caused by the interruption of the flow of oxygen to the blood or to the tissues by chemical damage.

The effect of asphyxiation is the autonomic shutdown of biological systems including respiratory, circulatory and finally neurological. PPE use will provide protection from most types of asphyxiation.

***Chemical:*** Chemical dangers are some of the most diverse and prevalent. The dangers to humans from chemical agents are in two main forms:

(a) **Toxic:** Causes long or short-term effects depending on concentration, exposure duration and route and type of toxin. An example of a chemical toxin is tear gas.

(b) **Corrosive:** Causes visible destruction or irreversible alterations in the skin tissue at the site of contact. An example of a corrosive chemical is hydrochloric acid.

Routine hazardous material precautions are the best protection.

***Etiological:*** Etiological agents are inclusive of biological agents. An etiological exposure occurs when a person comes in contact with another living organism contracting human disease. The form of this agent can be either viral or bacterial and usually present days to weeks after exposure. As with chemical hazards, routine HsAZMAT precautions are the best protection.

***Mechanical:*** Mechanical hazards are associated with explosive devices and can cause trauma through direct or indirect contact with physical (shrapnel) or mechanical (blast) hazards. Protection from this hazard can be obtained by:

(a) Delaying response or entry into the area.

(b) Limit time in hazard area by performing rapid entry for recon and rescue (limit exposure to dangers from secondary device)

(c) Remain under cover and at distance to hazard area (shielding, up-wind, up-hill).

## Preparation for the Incident Response

While many types of weapons are at the disposal of terrorists, statistics prove that bombs and incendiaries are still their number one weapon of choice. They are cheap (free if you steal the materials), easy to build, easy to conceal, and deliver to a target. Depending on the nature of the weapon, its level of destruction or method of use determines whether it is classified as a WMD. As outlined, WMD includes chemical agents, biological agents, and nuclear devices, radiological and high yield conventional explosives.

In preparing for any type of incident, each community should identify the most likely targets such as, place for public assemblies, court houses, the Post Office, Fire/Police departments, military establishments, schools/universities, power plants, laboratories, etc. Workers at these facilities should be acutely aware of anything or anyone appearing suspicious and report to the local law enforcement agency immediately.

They should also request an in-service from local law enforcement and fire fighting agencies on how to handle suspicious incidents and threats. Have an expert to suggest ways to make the likely targets less vulnerable to attack. Communicate with local law enforcement authorities and ask to be notified when they receive intelligence reports of any suspected terrorist activity.

Terrorists are likely not to give any warning before they strike. First Responders must stay vigilant. In such a situation, a cautious attitude forms the basis of good judgment.

## Response Actions

***General Actions*:** The actions required to be taken by the first responders are explained below.

***Protective Actions*:** These are those steps taken to preserve the health and safety of emergency responders and the public during an incident involving the release of WMD.

***Isolate Hazard Area and Deny*:** There is a need to keep everyone away from the area if they are not directly involved in emergency response operations. Unprotected emergency responders should not be allowed to enter the isolation zone. This 'isolation' task is done first to establish control over the area of operations, as well as provide room for operations, limit further exposures and finally ease accountability for responders, victims and the general public. This is the first step for any protective actions that may follow. Use the distances listed for the individual threat classifications.

***Evacuate***: People located in the threatened area are required to be moved to a safer area. To perform evacuation, there must be enough time for people to be warned, to get ready, and to leave the area. If there is enough time, evacuation is the best protective action. Begin by evacuating people nearby and those who are outdoors and in direct view / contact of the scene. When additional help arrives, expand the area to be evacuated downwind and crosswind to at least the extent recommended under the individual threat classifications. Even after people move to the distances recommended, they may not be completely safe from harm. The best recommendation is to send evacuees to a definitive place, by a specific route, far enough away so that they will not have to be moved again, if the wind shifts or the hazard area is expanded.

***In Place Protection***: People inside a building should remain inside until the danger passes. In-place protection is used when evacuating the public would cause greater danger than letting them stay where they are, or when an evacuation cannot be performed. People have to be warned /instructed to remain inside, to close all doors and windows and to shut off all ventilating, heating and cooling systems. In-place protection may not be the best option if:

(a) The hazard is an explosive, such as flammable liquids or gases, or when a secondary device is suspected.

(b) If it will take a long time for the hazard to clear the area; or

(c) If buildings cannot be closed tightly. Vehicles can offer some protection for a short period, if the windows are closed and the ventilating systems are shut off. Vehicles are not as effective as buildings for in-place protection though.

It is vital to maintain communications with competent persons inside the building so that they are advised about changing conditions. Individuals who are protected in-place should be warned to stay far from windows because of the danger from glass and projected metal fragments in a fire and/or explosion.

In order to further reduce confusion, a specific spokesperson should be identified both within the building and outside. Information should be provided to those protected in-place to ensure that they remain calm and informed. The incident commander or his designee should direct all communications.

## Response Phase

The initial response phase begins with the first report of a WMD incident to the local emergency services. The initial response element (IRE) should include local services such as firefighting, emergency medical and law enforcement agencies. The IRE will vary in capabilities from having no training in the response to terrorist incidents to a high level of training, equipment and resources capable of adequately performing the initial tasks for a successful outcome.

First responders will be the first to either identify the existence of a WMD terrorist incident, or become victims themselves. Therefore, it is necessary, for first responders to be cautious in their initial approach to any incident in which there is the possibility of a WMD threat. After the recognition of a WMD terrorist incident, it is necessary to implement protective measures and control the scene. These actions cover the fire and law enforcement duties to protect responders and the general public as well as preserve evidence for later investigative and prosecutorial measures.

As a WMD incident can range in size from small to extremely large in its effects and scope, there are four classifications of incident levels used as a basis for site management. The purpose of these levels is to provide others within the consequence management response system, at least to a degree, an idea of what to expect.

(a) ***Level I Incident***: It is routine emergency condition. These situations can usually be handled by first responders and require little outside resources in mitigating. Recovery time, if any at all, is very short. An example of a Level I incident could be a motor vehicle accident.

(b) ***Level II Incident***: It is a limited emergency condition, posing a potential threat to life and/or the environment. A small hazardous material spill such as gasoline from a storage tank could be an example of a Level II incident. Evacuation may be involved and more resources, such as special teams, may be required for successful mitigation.

(d) ***Level III Incident***: It is an incident that is considered a large-scale emergency condition. A Level III situation poses a major threat to life, property, and/or the environment. Evacuation is often necessary. A Level III incident could easily be an act of Terrorism involving WMD such as a building bombing or small-scale exposure to chemical or biological agents.

(d) ***Level IV Incident***: The final level, that is defined to be a full emergency condition. These situations pose major, long-term threats to life, property and/or the environment. Regional and sometimes national resources can be depleted in a relatively short period of time.

The first step in the actual response and mitigation is the isolation of the hazard area and the denial of entry. An Initial Isolation Zone (IIZ), which is defined as the area surrounding the incident in which persons may be exposed to dangerous and life threatening concentrations of material, should be established immediately. A Protective Action Zone (PAZ), which is the area downwind from the incident, in which persons may become incapacitated and unable to take protective actions, and/or incur serious or irreversible health effects, should be established as soon as possible.

The vicinity surrounding the affected area is also further divided into a Hot Zone, Warm Zone and Cold Zone. The Hot one is both the IIZ and PAZ combined. Any access to this area should be limited to emergency, personnel outfitted in appropriate individual protective equipment (IPE) to ensure their safety. The Warm Zone is for the decontamination of victims and those exiting the contaminated area. Incident Command System (ICS) personnel and command centers are usually located in the Cold Zone. At times, depending on the resources on-scene, staging areas can also be found in the Cold Zone. This area should be protected with an additional security perimeter to limit media and general public access.

During this initial response, the IRE should establish an ICS and designate and Incident Commander (IC); determine additional resources necessary to not only establish the IIZ, PAZ, and Hot/ Warm/Cold Zones, but also begin the identification of hazards involved, the treatment of victims, notification of proper authorities, and the beginning elements of the recovery and restoration phases within consequence management. These duties create a framework for the incident operation and make up what is commonly referred to as the Continued Response to an Incident.

Special rescue teams are oftenbrought in during the tail end of the first response phase along with support teams. These resources are many times is first break that first responders get, sometimes days after the incident has actually occurred. Many times the emergency has moved from the scene of the incident to tertiary care facility, such as hospital.

After the initial response and stabilisation of the incident, those leaders involved in consequence management should have a much more in-depth understanding of the situation and have long term plans in place. Understanding will include what type of device; chemical, biological, radiological, and/or nuclear was used and how to contain the affected area; limit further casualties and finally decontaminate and treat those affected. Determining the type and amount of resources needed to bring the incident to a close, and those resources having been requested should also be established. An estimated time period for mitigation and recovery should also be determined to judge what long-term resource allocation, financial and logistical support will be needed based on the length of the incident.

The term "mitigation" covers the initial response and continued response until recovery has begun. Once immediate response missions and life saving activities conclude, emergency teams are demobilised and the emphasis shifts from mitigation to recovery operations. The recovery phase is the second subset of consequence management. During this phase financial, administrative, logistic and operational supports are transferred to those assigned in recovery actions.

During recovery, significant resources can be used to restore the affected area depending on the damages sustained. A chemical device may take days to weeks, and necessitate extensive cleaning to be deemed safe. Biological devices are usually much more susceptible to the effects of the environment and may last only minutes, but are difficult to contain and clean, and, therefore, could remain in hibernation until re-released years later depending on the circumstances. The results of a biological toxin release can be long-term and spread amongst large numbers of the population. In case of, either a radiological or nuclear incident, there may be an extended recovery time from either radiation or conventional damages.

This phase involves areas of action such as rebuilding, decontaminating and/or the treatment of victims. Depending on the type of incident, this phase can last days to years while the infrastructure is re-established.

Every incident, especially acts of terrorism, is different. Each will have special problems and concerns. Actions to protect the public must be selected carefully. Actions narrated above help with initial decisions on how to protect the public at large. Officials must continue to gather information and monitor the situation until the threat is removed.

## Incident Specific Actions

***Bombs and Incendiaries*:** The terrorist goal is to create panic and mayhem. A popular method of doing so is with a secondary device. A secondary device is placed where responders and curious on lookers are likely to assemble and set to detonate with the aim to maim or kill unsuspecting bystanders and even the response personnel reacting to the primary device. A threat may even be received stating the bomb is in a specific location when in fact it is located in another place perhaps on another day, where people are most likely to congregate. The following steps should be taken in bombing incidents or threats:

(a) Evacuate to a distance of atleast 1,000 feet from the scene and set up an exclusionary zone even further out. Prohibit the use of radios, cell phones, pagers etc. any closer than 450 feet from the bomb.

(b) First Responders must be acutely aware of their surroundings and have a high degree of suspicion, as secondary devices, unexploded ordinance and or the terrorists may be present.

(c) Incident Commanders must take total control of the scene. Consideration for life safety issues, evidence preservation, exclusionary zones to keep unnecessary individuals out. Screen those being evacuated for possible contamination, injuries, eyewitness reports and identities. In case of suicide groups the culprits may be within the afflicted group. Law Enforcement personnel must be heavily involved with these tasks. A representative from all of the Public Agencies involved should be present in the command centre. The Command Centre should have proper communication, coordination and implementation of all resources involved must occur for a positive outcome.

(d) Areas for emergency mass casualty decontamination, responder decontamination and emergency evacuation should be established prior to patient contact and treatment. Large numbers of victims self-triaging themselves will quickly overwhelm responders and hospitals. The "Hot Zone" may have to be widened as the rescue and relief effort picks up and depending upon the number and type of casualties.

(e) For safety and security reasons, Fire & EMS personnel on the scene must be kept to a minimum, with strict accountability for every one on the scene. If entry must be made for rescue or for firefighting, it should be done in no less than full structural protective clothing with time in the blast zone kept to an absolute minimum. The "Swoop and Scoop" method should be used to retrieve live victims and the "Hit & Run" method of fire-fighting should be employed until an expert has declared the blast zone free of live ordinance. Trying to render normal patient care of firefighting tactics will put the responders and victims in mortal danger. Keep in mind that an explosive device could have been used to disperse a chemical, biological or nuclear material and normal firefighting gear will be virtually useless for these. Respiratory protection is vital.

(f) Document everything, take as many pictures and as much videography as possible and have everyone prepare a detailed statement of what they saw and did. Subsequently, the data collected can be analysed by the trained personnel.

(g) Use the proper chain of command in your community to report suspected terrorist incidents—sually progresses from local agencies to district administration to state Emergency Management Department (EMD) or the Governor/ Chief Minister who in turn will contact the central agencies. However, the information pertaining to the disaster/incident can also be passed through media channels.

***Chemical and Biological Agents:*** Large numbers of casualties when there are no injuries or physical signs of destruction which may or may not be accompanied by victim reports of peculiar odours, tastes and sensations suggest chemical or biological agents. Clues at the scene may also include containers with biohazard or poison labels, spray equipment and laboratory equipment that doesn't belong to anybody.

Chemicals such as nerve agents; blister agents, cyanides, chlorine, phosgene and riot control agents as well as certain biological toxins will have an immediate effect. Clues at the scene, signs and symptoms and victim reports may lead the responders to suspect that a chemical or biological agent is involved and allow one to institute timely precautions.

If a biological agent such as Botulinum, Anthrax, Plague, Salmonella, Small Pox, or any number of viruses are used without warning, the community won't know until it has potentially reached epidemic proportions. With Anthrax and Pneumonic Plague, once the symptoms have appeared,S death is usually inevitable. With an epidemic, local supplies of antibiotics and antidotes will be in short supply and local treatment centres will be overwhelmed, until state and central assistance is available.

Most agencies do not have the proper personal protective gear or the detection equipment for effectively dealing with chemical and biological agents. When the presence of chemical or biological agents is suspected, the scene must not be entered until it is rendered safe by qualified personnel. Hazardous material teams may have the proper detection and monitoring equipment and Individual Protective Equipment(IPE) to define the hazard. Otherwise, it maybe necessary to await the arrival of specialized teams. If entry into hazard area becomes absolutely necessary, it is essential to use the highest level of IPE available.

Any response to a chemical or biological attack will require the decontamination of equipment, entry personnel and casualties. This may be the single most important task facing the responders. The emergency decontamination of victims and survivors of a large chemical or biological attack will tax the abilities of any district or state. Pre–planning for mass decontamination needs to be accomplished as part of our preparation for terrorist attacks.

(a) Evacuate and isolate 1,500 ft in all directions.

(b) First Responder must be acutely aware of their surroundings and have a high degree of suspicion, as secondary devices, unexploded ordinance and or the terrorists may be present.

(c) Incident Commanders must take total control of the scene. Consideration for life safety issues, evidence preservation, exclusionary zones to keep unnecessary individuals out. Screen those being evacuated for possible contamination, injuries, eyewitness reports and identities.

Attempt to identify the agent. Law Enforcement personnel must be heavily involved with these tasks. A representative from all of the Public Agencies involved should be present in the Command Centre. The Command Centre should have proper communication, coordination and implementation of all resources involved must occur for a positive outcome.

(d) Areas for emergency mass casualty decontamination, responder decontamination and emergency evacuation should be established prior to patient contact and treatment. Large numbers of victims self-triaging themselves will quickly overwhelm responders and hospitals. The "hot zone" may have to be widened as the incident progresses. Anyone contaminated must not be allowed to leave the hot zone until the agent is identified and thoroughly decontaminated. Hospital medical staff should be brought to the scene to aid in triage and treatment.

(e) Decontamination solution can be prepared for nerve and blister agents, a .5% (10:1) water: bleach solution can be used on victims (do not rinse the eyes or open body cavity injuries with this solution). For equipment, use 5% bleach then wash with soapy water. For acute exposure to biological agents, decontaminate by scrubbing with soap and water, quarantine the victim(s) and give appropriate medical care. With irritants, scrub with soap and water and give appropriate medical care. Use the highest level of IPE available until the proper resources are available.

(f) If an entry is absolutely necessary, it is essential to use the highest level of IPE available.

(g) Use the proper chain of command in your community to report suspected terrorist incidents. This usually progresses from local agencies to district EMD to state EMD or the Governor/chief Minister who in turn calls the central agencies.

***Nuclear:*** While it is possible for a terrorist to acquire and detonate a nuclear device, it is highly unlikely. First responders would be more likely to encounter a device to disperse radioactive materials. This could take the form of bombing shipment of radioactive materials or a nuclear facility, or incorporating nuclear materials into the bomb.

The proper response to a bombing then should include a radiological monitoring device in addition to the other response issues outlined previously. This is not to say one sould enter the hot zone but merely to monitor the perimeters to establish the correct isolation distance.

## Incident Command System

The incident command system (ICS) is required to provide for effective operations at a multitude of emergencies and between different agencies with minimal interruption in normal operations, simple to adopt and learn by new users, adaptable to the diverse and ever changing-scope of emergency incidents, minor to major in extent; and share common and basic elements in organization, terminology, and procedure.

The key concept in the ICS system is the span of control. A hierarchy of command and breakdown of duties inherent in an emergency is required to be developed to limit the span of control, or supervisory extent of officers to few individuals. The system is set up to manage the laid down number of people with this span of control but is completely expandable and collapsible based on the individual incident, and its corresponding requirements. Each of the levels is explained below.

Command responsibilities are at an executive level and are managerial in nature. They include establishing objectives, insuring the incident progression appropriate for those objectives, the priority of operations, and flow of information to the media as well as other ICS participants. The Incident Commander (IC) is the overall commander of the incident and is supported by a command staff that reports directly to the IC.

Operations Sector responsibilities are of line nature, or the workhouse of the ICS system. They include the operations involved in fulfilling the command objectives, and providing information on all aspects of the incident to designated command staff member. This sector is the most widely used in the ICS.

Planning Sector operations are of a staff or executive in nature. The responsibilities of this sector include the analysis of information to develop plans of action according to the resource status, situation data and future predictions. This position is especially necessary in lengthy operations to ensure smooth operations and long-term endurance.

Logistics Sector responsibilities are of a staff nature as well. This sector ensures that personnel, equipment and services required to maintain the incident are available to the IC.

Finance Sector responsibilities are also of a staff nature. This sector provides for financial management and accountability. The Finance sector again is for long-term operations and supports the Logistics Sector in obtaining resources.

Under each of the sector sections, various levels of further breakdown are applied based on the incident scope. It is highly recommended that all responders take the time to understand the ICS system through further study. When used appropriately, the ICS can alleviate significant confusion and allow for a more effective and efficient operation.

## Mass Casualty Incident

A Mass Casualty Incident (MCI) is any incident that taxes the resources of those on the scene, or responding. The incident could be a personal injury accident with multiple patients or a building collapse with hundreds of victims. Therefore most if not all, WMD incidents will be classified as MCIs. Jurisdictions should have established MCI contingency plans to operate under these conditions. These same plans can be easily adapted to WMD incidents.

The major concept of the MCI is triage, or the quick determination of patient acuity and treatment classification. This process allows for the quick allocation of resources and processing of high numbers of patients with the resources available. Once triage is accomplished, a visible tag or identification is carried with each patient allowing those in follow-up treatment stages to be aware of triaged acuity and injuries of the patient. This allows those who encounter the patient following the initial triage to bypass the primary assessment for more effective treatment.

Systems of triage vary by jurisdiction and should be covered within pre-existing SOPs, though there are four universal classifications of triage acuity, each outlined as given in Table 6.1.

**Table 6.1 Triage Acuity**

| Classification | Meaning | Injuries such as |
|---|---|---|
| Green | No hospital treatment needed; first aid only; care needed in hours to days | Walking wounded; minor cuts, scraps with controlled bleeding; fractures of small bones; other injuries requiring limited non-emergency medical attention. |
| Yellow | Non-urgent; hospital care needed within hours | Altered level of consciousness; fractures of long bones; uncontrolled bleeding; other injuries requiring medical attention. |
| Red | Urgent; hospital care; care needed within minutess | Unconsciousness; signs of shock; fractures of femur/ pelvis, severe uncontrolled bleeding; respiratory arrest with pulse; other injuries requiring emergency attention with probability of survivability. |
| Black | Deceased or unsalvageable; no CPR; no care needed | Injuries incompatible with life such as cardiac arrest; exposed brain matter; decapitation, etc. |

Source: *Weapons of Mass Destruction (WMD) Handbook*, U.S. Department of Defence, Washington, D.C., February 2001, p. 42.

Other than patient triage, the key to the successful outcome of a MCI is system status management. This includes the effective and efficient use of resources including hospitals, ambulances, medical personnel and other transport vehicles (e.g. buses, ambulances).

Any department /agency's current MCI preparations or major incident facets can be easily adapted to cover most WMD incidents. The differentiating factor that needs attention is patient decontamination in many cases. This aspect of an MCI requires additional resources and time that must be planned for in advance.

As with all operations, training, including tabletop and field exercises, be conducted routinely to ensure that all responders and personnel are familiar with MCI operations. By doing so the plan can be adapted according to any particularities and plan deficiencies can be identified.

The first responders are the most important elements towards effectively managing the consequences of a Weapons of Mass Destruction incident. Regular practice by all the elements forming part of the Response Team will pay rich dividends in terms of saving precious human lives. These responders should also be suitably equipped so that they themselves do not become casualties. Preparation for and building up the capability to deal with the consequences is itself a deterrence.

# 7

# Specialised Response Assets

*The threat of nuclear terrorism is for us very dangerous. The fact is that on the globe several tons of enriched uranium has disappeared from the sites where it was produced and stored.*

**Vladimir Kryuchkov**
**(Ex KGB Director)**

The current trend in terrorism is evolving towards more destructive means, especially towards employing weapons of mass destruction. This evolution is occurring for a number of reasons. Firstly, the disintegration of the Soviet Union and displaced WMD, secondly the evolving terrorist motivations and emerging ideologies, thirdly the proliferation of technologies of mass destruction and increased access to information and information technologies, all together have increased the vulnerability to terrorist attack. Terrorists groups, non-state and state sponsored, violence appears to be moving towards being more deadly and destructive. Spanning the last decade, terrorist incidents that have potential for large-scale casualties are occurring more frequently and are accruing higher casualties. The reason for rise in terrorism may be "because there are increasing number of violent non-state actors for whom the logic of limited lethality applies only slightly".[1]

NBC terrorism may be anticipated from fanatical religious groups and cults, Islamic extremists, right-wing chauvinists and loosely affiliated terrorists who have the traditional concern or desire for group preservation. The proliferation of WMD munitions and technology from or to an undesirable nation is an additional source of this growing terrorist threat. Most of the developed nations have evolved strategies and raised specialised response assets to deal with the crisis and consequence management of WMD terrorism.

## United States

The United States of America in 1975 had recognised the threat of terrorists using Improvised Nuclear Device (IND) within US and had accordingly created Nuclear Emergency Search Teams (NEST) under the Department of Energy. The NEST provides technical response to resolution of incidents involving improvised nuclear and radiological dispersal devices. The team is capable of searching, locating and identifying nuclear and radiological devices or materials. However in 1995, the United States earnestly started preparing to combat the threat of weapons of mass destruction (WMD) terrorism within US following the release of sarin gas on a Tokyo subway by the Aum Shinrikyo, a Japanese cult group and the Oklahoma City bombing with the signing of Presidential Decision Directive 39(PDD-39) by President Bill Clinton, which addressed the way United States would prepare itself to deal with the terrorist use of WMD. As per the Directive, Federal Bureau of Investigation (FBI) is the Lead Federal Agency for crisis management for domestic incident and Department of State for foreign incident. For Consequence Management, the Federal Emergency Management Agency (FEMA) is the Lead Federal Agency (LFA) for domestic incidents and Department of State for foreign incidents. The Federal Response Plan has been prepared in April 1999, which gives out the specific responsibilities to various federal and state agencies covering the entire domain of consequence management of a WMD incident.

In order to deal effectively with a WMD incident, unique consequence management assets have been created under the Department of Defence (DoD) in the United States.

***Army Technical Escort Unit*:** The Technical Escort Unit (TEU) was established in 1944 and is the Army's oldest chemical unit. Its missions include worldwide response for escorting, rendering safe, disposing of, sample verification, mitigating hazards and identifying weaponised and non-weaponised chemical, biological and hazardous material. Military and civilian personnel possess a wide variety of specialised training in explosive ordnance identification and handling, radiography, military and commercial handling, chemical and biological detection and monitoring equipment. Deployment packages include protective equipment, hazardous material transfer systems, mobile systems for detecting, monitoring and identifying chemical and biological agents and communication links.

TEU would be deployed as part of the crisis management effort. Ideally, a device will be located and can be transported before it is employed. TEU provides the capability to transport an agent or render a device unusable or ineffective.

***Chemical Biological Incident Response Force***: The United States Marine Corps Chemical Biological Incident Response Force (CBIRF) is relatively a new Marine Corps unit, created in the wake of 1995 sarin gas attack on Tokyo's subway system and is designed to provide emergency support following WMD terrorist incidents. The concept for employment of the CBIRF details an initial, rapid response to chemical or biological incidents. When such an incident occurs, the CBIRF will deploy to the affected site. Once there, the CBIRF will provide a number of significant initial consequence management capabilities, assistance in coordinating initial relief efforts, security and isolation at the effected site (when authorised), detection, identification and limited decontamination of personnel and equipment, expert medical advice and assistance, and service support assistance. Throughout its response, civilian and government consultants in areas related to chemical or biological incidents will advise the CBIRF.

***Rapid Assessment and Initial Detection Teams***: The initial military response for consequence management after an incident is likely to come from National Guard Rapid Assessment and Initial Detection (RAID) team. The team consists of 22 representatives from a cross section of functional areas that can deploy and assess the situation, advise the local, state and federal response elements, define requirements and expedite employment of state and federal military support. Their mission is to provide early assessment, initial detection and technical advice to the incident commander, during a WMD incident and identify the requirements for DoD support.

The RAID teams can rapidly deploy to an incident site and provide initial support to the Incident Commander. The element has the ability to conduct reconnaissance, provide medical advice and assistance, perform detection, assessment and hazard prediction, and can provide technical advice concerning WMD incidents and agents. RAID teams will also have a significant reach-back capability that allows them to tap into expertise across the country. A RAID team is organized under the peacetime control of a state's Adjutant General.

**Nuclear/Radiological 52nd Ordnance Group (EOD).**

The 52nd Ordnance Group (EOD) is the only active duty Army Explosive Ordnance Disposal Group. Assigned to Forces Command, it has operational and administrative command of four EOD Battalions, each of which has ten companies. The units of this group provide explosive ordnance disposal units to defeat or mitigate the hazards from conventional, nuclear or chemical military ammunitions and weapons of mass destruction devices. This capability compliments the capabilities provided by the Department of Energy's Nuclear Emergency Search Team (NEST). Members can also work with explosive components of chemical and biological weapons. It also provides continuous support to the US Secret Service and State Department for VIP protection details. Each unit has a variety of bomb disposal tools and detailed classified procedures for handling US, foreign and terrorist munitions. Included in their equipments are robots for remote operations, special cannons and explosive shape charges and a variety of EOD tool sets for specific munitions.

***WMD Civil Support Teams*:** WMD Civil Support Teams are full time National Guard teams that, when certified, provide assessment, analysis and communications to identify and support follow-on federal/ state response assets. The team carries out assessments of a suspected NBC or radiological event, advise civilian responders regarding appropriate action and facilitates request for assistance to expedite arrival of additional state and federal assets to help save lives, prevent human suffering and mitigate property damage.

## Chemical Biological Rapid Response Team (CB-RRT)

The CB-RRT is the Department of Defence (DoD) joint national response asset, which on orders deploys and establishes an integrated capability to coordinate and synchronise technical assistance (medical and non medical) to effectively respond in both the Crisis and Consequence Management to a WMD incident/ terrorist attack. The CB-RRT provides a technical support package specifically tailored for response requirements. Upon arrival at an incident site, the CB-RRT deploys command elements, quickly establishes initial coordination with the Lead Federal Agency and prepares to deploy an advisory team to the federal, state and local command and control organisations as required or directed by the designated operational commander. It also coordinates and plans assistance to local authorities and first responders for consequence management operations. The CB-RRT can also bring together some of the nation's leading experts in chemical and biological WMD issues without the need for them to be deployed to the incident's site. The CB-RRT organises, based on the situation, to provide the appropriate level of graduated response and technical expertise necessary to assist in mitigating a chemical or biological incident.

***Medical Research Assets***: There are a number of Department of Defence (DoD) medical research institutes and military laboratories, which are designed to study WMD and their effects. The US Army Medical Research Institute for Infectious Diseases (USAMRIID), for instance, would work closely with Centre for Disease Control (CDC) in the event of a biological terrorism event. USAMRIID conducts research to develop strategies, products, information procedures and training programmes for medical defence against biological warfare threats and infectious diseases. It develops products such as vaccines, drugs, diagnostic tests and medical management procedures to protect military personnel. It also serves as the reference centre for identification of biological agents from clinical specimens and other sources. Other laboratories such as Army Medical Research Institute for Chemical Defence (USAMRICD), and Armed Forces Radiobiological Research Institute (AFRRI) focus on chemical and nuclear weapons and their effects on human beings respectively.

## North Atlantic Treaty Organisation (NATO)

In order to meet the changing security threats including use of weapons of mass destruction (WMD) by terrorists on European soil, NATO in November 2002 agreed to improve capabilities in a wide range of areas, including nuclear, biological and chemical weapons defence initiative.[2] The Czech Republic on behalf of NATO has raised a multinational Chemical, Biological, Radiological and Nuclear (CBRN) Defence Battalion, which offers support to NATO Response Force (NRF). The battalion became fully operational on 01 July 2004. Although Czech –led, it has eleven other contributing nations, and as a framework nation it is the largest contributor of personnel and equipment. Such a level of diversity in a single battalionis almost unthinkable in other functional areas of combat. However, the complexity of the CBRN environment is such that, it is more important to have genuine expertise to advise commanders and manage the consequences. The battalion brings together specialist capabilities from twelve nations and all the NATO members have invested in the battalion by providing state of the art capabilities and has the provision of reach-back capability to enhance the analysis.

The role of the battalion is to conduct CBRN recce, detection, identification, surveillance and decontamination. The battalion is organised into three task forces each capable of CBRN recce and decontamination on task-organised basis. It has a BW detection company provided by UK, which is capable of sample collection, sample analysis and identification of a number of BW agents including anthrax and botulinum toxin. The Czech provides the lead for chemical and radiological laboratory, which has the capacity to collect samples and identify them. The battalion also contains experts in NBC Explosive Ordnance Disposal, a medical detachment that provides doctors who specialise in CBRN effects, mainly in consequent management during decontamination operations and five platoons to decontaminate personnel, vehicles and infrastructure.

The proliferation of WMD is widely recognised as the most serious threat to the global security situation. Nations are preparing to deal with the threat of terrorists using WMD in their territories. India is developing to become a major global player in world affairs and accordingly needs to prepare to deal with this threat by having a comprehensive strategy and necessary assets.

## Specialised Response Assets For India

Chemical and biological weapons are likely to be the first choice amongst the WMD to be used by the terrorists/non-state actors because of their easy availability as also due to their potential for enormous, grievous, collateral damage. The weapons produce immediate and delayed effects that hamper operations through the contamination of individuals, equipment, supplies and critical areas.

In order to manage the consequences of WMD by terrorists there is a need to have a specialised force, which is capable of agent detection and identification, casualty search, rescue, personnel decontamination and emergency medical care.

**Chemical Biological Incident Response Force (CBIRF)** of the United States of America is a national consequence management asset capable of rapid response to chemical or biological attacks initiated by terrorists. An integral part of the nation's nascent domestic consequence management preparation, CBIRF has been regularly pre-positioned for such potential terrorist targets as 1996 Summer Olympics in Atlanta, the Presidential Inauguration and for other high profile events. CBIRF actively participated in dealing with the Anthrax attacks in US following the 11 September attacks by terrorists. CBIRF participated in Joint Senate offices sample collection operations in which the team members collected over 600 samples and participated in joint removal and packaging of approximately 10,00,000 pieces weighing 12 tons of unopened/suspicious mail.

CBIRF was the vision of the US Marine Corps Commandant, General Charles C. Krulak. Following the March 1995 Tokyo sarin gas attack, General Krulak then Marine Forces Pacific commander, went to his commander and asked what assets Pacific Command could muster to assist the Japanese. The answer he got: "We've got nothing".[3] Upon becoming Commandant in July 1995, General Krulak, with no tasking from Department of Defence established the CBIRF. By April 1996, the CBIRF vision had become an operational force, created out of existing Marine Corps personnel, infrastructure and budget.

In and of themselves, the elements that make up CBIRF are routinely military. Together however, they are truly a revolutionary organization. CBIRF includes reconnaissance, detection, decontamination, medical, security and service support elements. With these capabilities "under one tent", CBIRF has the ability to provide command and control support to civilian incident site commander, conduct detection in a contaminated environment, insert Navy doctors into a contaminated ("hot") zone for immediate triage of the more severe cases, decontaminate the victims and turn the victims over to the local health authorities outside the hot zone. The unit also has memoranda of understanding with numerous government agencies such as the Naval Medical Research Institute. The force is additionally supported by an "electronic reach back" group of scientific and medical consultants headed by Nobel Laureate Joshua Lederberg. Nowhere else are such assets contained in one deployable unit and no other organisation has developed a comparable extended support structure ready to help in a crisis. With its elements cross-trained in their respective specialties, CBIRF has become the tactical touchstone of consequence management: turning contaminated victims into treatable patients.

Besides the command element, headquarters and administrative support company, the teeth of the unit are contained in the Reaction Force Company, which is 170 strong and comprises of reconnaissance, technical rescue, extraction and decontamination platoons.

***Concept of Operations***: The concept of operations of the CBIRF is based on two stages of responses, driven by time. An Initial Response Force (IRF) composed of 123 personnel is kept at "One Hour" notice 365 days a year. The equipment, wherewithal and training of this force enables it to make two entries into the contaminated environment. The Follow-on-Force (FOF) follows up the IRF. The FOF consists of 200 personnel and is maintained at four hour notice.

The IRF is capable of carrying out detection/identification of CBRNE hazards and can perform rapid extraction of casualties. The equipment with IRF enables it to decontaminate 65-75 non-ambulatory and 200-225 ambulatory casualties per hour. It can also provide immediate relief and rescue, medical triage, stabilisation and medical regulation. IRF has experts for disposal (make safe/destroy/diffuse/deactivate) of explosive ordnance against time. IRF is a self-sustaining force with complete wherewithals for command, control, communications and logistics.

As the name suggests, FOF is a second echelon response force that follows up on the IRF. This 200–strong force is capable of detection, decontamination and re-supply as required. This force is capable of deployment from APOE (Airport of Embarkation) in eight hours. IRF and FOF make one composite entity embodying the total capability claimed by CBIRF.

The IRF-FOF combine have been equipped and trained in a manner so as to have wide-ranging operational capabilities covering all aspects of consequence management.

***Mission***: When directed, forward-deploy and /or respond to a credible threat of a Chemical, Biological, Radiological, Nuclear or High Yield Explosive (CBRNE) incident in order to assist district, state or central agencies in the conduct of consequence management operations by providing capabilities for agent detection and identification, casualty search, rescue and personal decontamination and emergency medical care and stabilisation of contaminated personnel.

***Employment***: The force is to be employed to support the lead central agency in coordinated emergency management measures to protect public health and safety, restore essential government services and provide emergency relief to individuals and businesses affected by the consequences of terrorism.

## Operational Capabilities

The IRF-FOF combine have been equipped and trained in a such manner so as to have a wide ranging operational capabilities covering all aspects of consequence management. Salient highlights of the operational capabilities are as under: -

(a) ***NBC Reconnaissance***: The wide variety of detection equipment held permits detection and identification of 120,000 Toxic Industrial Chemicals (TICs), as well as all known chemical and biological agents. It is also capable of collecting biosamples for lab analysis and detecting and measuring Alpha, Beta, Gamma, Neutron and X-Ray Emission.

(b) ***Casualty Extraction***: The Rapid Extraction Team can extract 6-12 non-ambulatory casualties per hour. The extraction capabilities are distributed over three levels of difficulties; accordingly the search and extraction teams are constituted at three levels A, B and C.

(c) ***Decontamination***: The IRF carries equipment for setting up field decontamination station. This provides the capability of decontaminating 200-225 ambulatory casualties and 65-75 non-ambulatory casualties per hour.

(d) ***Technical Rescue:*** The qualification of the word 'rescue' with 'technical' is indicative of the fact that this rescue has a high degree of difficulty and requires specialist technical equipment and skills. Special features include confined space rescue, trench rescue and rescue using advance rope techniques, re-stabilising the collapsed structures and capability to search out victims trapped under collapsed structures.

(e) ***Explosive Ordnance Disposal (EOD):*** In the field of explosive ordnance disposal, extensive use of Robotics has been made. Robots provide the capability of unmanned reconnaissance, portable X-Ray with real time imagery; detection/measure Alpha, Beta and Gamma radiations and IED render safe capabilities.

(f) ***$C^4$ I Support***: A wide cross-section of communication equipment based on secure voice/data using tactical/

civilian SATCOM as the media provides C4I support both within the various components of the IRF as well as between the IRF outside agencies.

(g) ***Logistics and Medical Support***: The unit is adequately equipped for logistics and medical support. It has organic transport, supply lines and stocks of medicine and medical experts to provide support for approximately 150 trauma patients with a range of injuries, antidotes for 1500 nerve agent casualties and wherewithal for airway medical management (respiratory distress cases).

Chemical Biological Incident Force (CBIRF) of the United States of America provides a number of significant initial consequence management capabilities and is therefore, felt that a similar kind of specialised unit be raised in India under the aegis of the Indian Army and form part of the Special Forces.

## Nuclear Emergency Teams

Although the probability of nuclear terrorism is very low, its horrendous consequences demand continuous vigilance. A terrorist nuclear attack need not necessarily involve a nuclear explosion. Nuclear power plants cannot withstand attack by a large jumbo jet full of fuel. There is always a possibility of terrorists stealing nuclear material and preparing a crude radiological device, which would not cause a nuclear explosion, but would scatter radioactive material over a populated area causing panic and terror.

In order to deal with such contingencies there is a need to have specially trained teams, which are capable of operating in a nuclear or radiologically contaminated environment. These Nuclear Emergency Teams should be able to carry out following tasks:

(a) Search and identify radioactive source.

(b) Continuous monitor radioactivity level at the site of incident.

(c) Capable of operating in NBC contaminated environment in full individual protective equipment.

(d) Capable of carrying out prediction of areas likely to be contaminated due to radiological release.

(e) Advise local administration the steps to be taken to save lives.

(f) Should be capable of reaching the incident site in the shortest possible time.

(g) Should be capable and trained to operate with the Special Forces.

**Employment**

Nuclear Emergency Teams (NET) should be employed on the order of the Crises Management Group (CMG). They should be raised and trained by the Department of Atomic Energy. Regular exercises should be carried out with the Special Forces of the Army and National Security Guards (NSG) to augment and improve their efficiency. They should preferably be located in major metros.

**Chemical Emergency Team**

In addition to the Chemical and Biological Incident Response Force (CBIRF), there is a need to raise Chemical Emergency Teams (CETs) which should be specifically raised, equipped and trained to handle terrorists chemical attacks and disasters. The CETs should be capable of carrying out following tasks: -

(a) Search and identify chemical agent source.

(b) Capable of operating a NBC contaminated environment in full individual protective equipment.

(c) Capable of monitoring the chemical contamination levels in an area.

(d) Capable of carrying out prediction of areas likely to be contaminated due to chemical cloud.

(e) Carry out decontamination of casualties at the site of the incident and provide first aid.

(f) Provide advice to local hospitals on measures required for treatment of survivors.

Chemical Emergency Teams (CETs) should be employed on the orders of the Crisis Management Group (CMG) and should be raised under the aegis of Ministry of Environment and Forests. They should regularly train with the other such teams including NSG. They should also preferably be located in all major metros.

## Biological Emergency Teams

The likelihood that biological weapons will be used against nation states continues to increase. Despite international convention (BTWC, 1972), banning biological weapons, several countries still possess them. In their destructive power, the biological weapons are no less than with nuclear weapons, if they are used keeping weather and other factors in view. With the advances made in biotechnology, the production of such weapons by the terrorists groups has become a reality.

A bio attack is very different from a nuclear or chemical attack, since it is usually difficult to pinpoint the time and place of the attack in their case. The effects of such an attack on the targeted population takes time and spread very fast. This is likely to lead the public health system under tremendous strain and is also likely to lead law and order crisis since panic would spread in the population.

Biological Emergency Teams (BETs) sould be one organisation, which would be required to be trained and equipped to handle such incidents. BETs should preferably have the following capabilities: -

(a) BETs would be the first responders in case of biological attack.

(b) Provide technical advice and help control situation in conjunction with state and local administration.

(c) Carry out medical examination of the affected population.

(d) Help the hospital authorities in antibiotic therapy, immunisation, mass vaccination and other public health related measures.

(e) Help in quarantining of the affected population/ hospitals.

(f) Provide regular feedback to National Centre for Communicable Diseases (NCCD) and the Control Room.

Biological Emergency Teams (BETs) would be employed on the direction of CMG by National Centre for Communicable Diseases (NCCD). These teams would comprise of specialist doctors to include physicians, pathologists, epidemiologists, microbiologists and other specialists as required. Each team should be located at each mega city. These teams should be ready to respond to any emergency from any part of the country and should be trained and equipped with suitable diagnostic equipment to detect viruses and bacteria and deal with biological emergencies. The responsibility for raising of these teams would be with that of the Ministry of Health and Family Planning in consultation with Ministries of Agriculture and Environment Forests.

## National Centre for Communicable and Infectious Diseases (NCCID)

The existing National Institute of Communicable Diseases (NICD) should serve, as the National Centre for Communicable and Infectious Diseases (NCCID) established at New Delhi. The centre should have facilities for testing and research on all aspects of Biological Warfare agents. In addition, there is need to have Regional Centres, at mega cities known as Regional Centre for Communicable and Infectious Disease (RCCID). The main function of NCCID and RCCID could be: -

(a) NCCID should be responsible for all research and related activities to support a concerted biodefence effort to cater for Biological warfare possibilities.

(b) Maintenance of cultures and viruses and bacteria under secure conditions.

(c) Mobile facilities for testing and diagnosis of samples for virus, bacteria and other pathogens.

(d) Coordinate activities of various RCCIDs.

(e) Maintenance of Biological Emergency Teams (BETs) for dealing with biological emergencies onsite.

(f) Coordinate and maintain health intelligence of unnatural diseases that outbreaks throughout the country.

(g) Coordinate mustering up of all vaccine and drug production agencies in the country for production of large quantities as and when required.

(h) Networking with other institutions including hospitals both nationally and internationally.

(j) Carryout Research and Development activities.

The above-mentioned specialised response assets have been identified at the Centre Similar response assets would be required at the state and district/metro level depending upon the likely threat. There would be a need to provide additional funds for raising, training and equipping of those organisations. Help from friendly foreign countries for the purpose of training and equipping these forces should be explored. We must take up the yoke of preventing such attacks and prepare for consequence management – managing the aftermath of such attacks – with vigour.

## References

1. R Falkenrath and B Thayer *NBC Terrorism and Covert Attack,* (Cambridge: The MIT Press, 1998).
2. Roger Lane, *NATO Transformation: The Development of a CBRN Defence Capability, RUSI Journal,* August 2004, pp.40-41.
3. Chris Seiple, *Consequence Management: Domestic Response to Weapons of Mass Destruction, Parameters,* Autumn 1997, p.122.
4. Ibid, p.122.

# 8

# Role of Armed Forces

*If I always appear prepared, it is because before entering an undertaking, I have meditated for long and foreseen what may occur.*

**Napoleon**

By virtue of their inherent strength, motivation and discipline, the Armed Forces would be required to play an important role while managing the consequence of Chemical, Biological, Radiological, Nuclear and High Yield Explosive (CBRNE) terrorist incident. They are perhaps the best organized and managed to provide support to a full range of public services such as communications, transport, emergency medical care, search and rescue, engineering and other support activities. They are capable of reacting quickly and respond rapidly in a self–contained, self-sufficient manner and with mobility. The personnel are well trained in individual skills necessary to perform their professional and functional activities and can function under an integrated / flexible management system. The Armed Forces have an enormous inherent potential to serve as an additional instrument for effective response and delivery of emergency assistance.

However, experience has shown that if the armed forces are to be fully effective in this context, their capabilities must be fully understood by relevant counter-disaster authorities; also their roles and responsibilities must be clearly defined in counter WMD incident/disaster plans. The organisation and management systems of armed forces make them suited for operations under disaster conditions. In addition, many of their normal activities are parallel to those in public emergency services. Thus, they can offer invaluable support in engineering, communications, transport, rescue, emergency medical services and water supply. The units available to provide these services are usually deployed throughout the country and close to centres of population, where disaster may have its most significant effect. These military forces have quick reaction capability and can respond rapidly, taking with them their own management, communications and administrative systems in a self-contained, self sufficient and highly mobile form. They are well trained in the individual skills necessary to perform professional and functional activities and practice in collaborated and coordinated action under their own flexible management systems. Add to this the capability for sustained operations away from their home base in all kinds of weathers by day and night, and one sees clearly the potential inherent in any military organization for effective consequence management of WMD terrorism.

On the other hand, the prevailing threat perception in the region also point towards the possibility of the armed forces and their installation coming under attack by anti national elements terrorists, using Weapon of Mass Destruction (WMD), whereby they would be required to respond quickly and decisively. The armed forces would be required to prepare themselves to deal effectively with this new and potent threat. The armed forces would be required to develop doctrines for domestic support in case of WMD incident/ disaster in the country or in our neighbourhood and operations in NBC environment including a WMD strike on its installations.

The primary goal of Indian counter terrorism programme is deterrence. A strategy that makes the most effective use of available capabilities will allow India to better counter the strategies of those groups that use weapons of mass destruction. The Indian Armed Forces offer many an important tool for use in such a strategy.

## Consequence Management

The role of the Armed Forces in countering WMD terrorism is one of strategic support. Military forces are primarily designed to operate against threats from outside the country. Many of the skills required for combat, however, are applicable too in domestic WMD emergencies and can supplement the capabilities of local, state and central agencies responsible for managing a crisis and its consequences. In many cases, an incident involving WMD will be of such scale so as to exceed the resources of other agencies. The Armed Forces have therefore, been tasked through legislation to provide aid to civil authorities which includes preparation for, resolution of, and consequence management after a terrorist incident/disaster. The civil administration, the local police, the medical authorities etc. must be as well trained and equipped as the armed forces to take on the task of consequence management.

A number of agencies at the centre, state, district and local level will participate in a WMD terrorism effort. An understanding of their requirements and expectations will enable the military to provide the most effective support. However, to obtain maximum benefit, it is mandatory for these agencies to develop an effective relationship with the local units beforehand, and include local military personnel in the planning process.

The armed forces may be called upon to support the district and state disaster management agencies immediately in the event of a terrorist attack. The forces, immediate response may include assistance to civil agencies in meeting the following types of emergancies:

(a) Rescue, evacuation and emergency medical treatment of casualties, maintenance or restoration of emergency medical capabilities and safeguarding the public health.

(b) Emergency restoration of essential public services (including firefighting, water, communications, transportation and power).

(c) Emergency clearance of debris, rubble and explosive ordnance from public facilities and other areas to permit rescue or movement of people and restoration of essential services.

(d) Monitoring and decontaminating radiological, chemical and biological effects, controlling contaminated areas, and reporting through national warning and hazard control systems.

(e) Roadways movement control and planning.

(f) Safeguarding, collecting and distributing food, essential supplies and material on the basis of critical priorities.

(g) Damage assessment.

(h) Interim emergency communications.

(j) Facilitating re-establishment of civil government functions.

## Aid to Civil Authorities

The Armed Forces support to domestic consequence management of WMD incident is a contingency mission in response to request by civil administration. The support is likely to be short-term, designed to transition to civilian agencies as soon as possible and the return armed forces to their assigned roles.

The support focuses on providing specialised assistance and is designed to bring the situation to a point where the civil administration and other agencies can fully manage affairs. The support is contingent on the availability of the forces without constraining actual operational role and affecting their war fighting capability and posture. The support is also contingent on the risk and time / distance factor. However, the response will be immediate with the aim of saving maximum possible lives and property and mitigate human suffering.

**Legal Provision:** The aid to civil authorities in case of a CBRNE incident will continue to be governed by the provisions as provided in the constitutional and the legal framework for the armed forces to render assistance during all types of disasters and calamities when the situation is beyond the capability of local civil administration. *"Aid to Civil Authorities by the Armed Forces"*, a Ministry of Defence Pamphlet dated 30 Nov 1970, provides necessary sanction for deployment of Armed Forces in aid to civil authorities.

**Type of Assistance:** The aid by the Armed Forces will be requisitioned mostly in a reactive scenario, i.e. when the incident has already taken place. It involves the following:

(a) ***Command and Control***: The Armed Forces are capable of providing infrastructure for setting up of command and control organisation for providing relief. It also includes co-ordination support in the form of restoring command and control by the affected agency by providing survey, assessments and reconnaissance.

(b) ***Medical Aid***: Provision of medical care with the help of medical teams including treatment in the nearest Armed Forces Hospital. Subsequently provide medical care at the site of the incident by establishing Mobile Field Hospital. Carry out triage / treatment and decontamination of the affected population. Provide teams of doctors and other technical personnel to augment the civil hospital facilities.

(c) ***Transportation***: Provision of transport to include aircraft / helicopters/ ships /boats / vehicles for transportation of relief material to the affected areas and also for moving out casualties for further treatment.

(d) ***Engineering***: The engineering support would be required in terms of power generating, demolitions, earth moving, road repairs and camp constructions.

(e) ***Communications***: Provision of communications from the site of the incident to outside, since, certain facilities would have been destroyed or might have been contaminated. Provision of communications to the outside world will result in better management of relief requirement to tide over the calamity. Effective communications will facilitate right type of relief in terms of trained manpower, equipment, medical supplies reaching the site, which will facilitate early recovery.

(f) ***Civil Affairs:*** The Armed Forces can help the civil administration in setting up and running of relief camps, repair, maintenance and running of essential services like electricity, water supply and also provision of food to the affected population.

(g) ***NBC Reconnaissance:*** The specially trained and equipped troops will be able to advice the civil administration on evacuation /rescue of population and restrict the hazard, detect, identify and constantly monitor chemical / biological agents and radiation stage.

(h) ***Search and Rescue***: The Armed Forces with the specialised response assets will be able to carry out search and rescue operations at the site of the incident.

## Principles of Employment of Armed forces

The operations of the Armed Forces, whenever called upon to assist the civil authorities in dealing with WMD incident/ disaster are governed by the following guidelines:

(a) ***Command of Troops***: The specialised response assets from the Armed Forces whilst operating for dealing with WMD incident/disaster should continue to be under command of their own commanders.

(b) ***Requisition of Aid on Task Bases***: Requisition of aid should not be in terms of number of troops, engineers and medical team. Civil administration should instead spell out tasks and leave it the Joint Task Force commander to decide on the force level, equipment and the methodology.

(c) ***No Menial Task***: The casualties in a WMD incident are likely to be heavy, but it must be remembered that forces employed for consequence management are not employed for menial tasks. These task need to be performed under the aegis of civil administration / NGOs.

(d) ***Regular Liaison and Co-ordination***: In order to ensure that optimum benefit is derived, regular liaison and co-ordination needs to be ensured at all levels and contingency plans made and disseminated to the lowest levels of civil administration.

(e) ***Advance Planning and Training***: The Armed Forces while planning to deal with incidents related to WMD will have to cater for all types of contingencies which they are likely to encounter. They will have to carry out regular training with other forces like police, fire department and medical personnel from civil hospitals who are going to be the first responders. Regular exercise and war games will have to be planned and carried out to stay ahead of the terrorists who are likely to use WMD.

(f) ***Integration of all Available Resources***: All available resources, equipment, accommodation and medical resources with civil administration, civil firms, NGO's etc. need to be taken into account while evolving plans to deal with WMD incidents. All the resources need to be integrated to achieve optimum results. Similarly, all assets of the Armed Forces must be integrated to obviate duplication or overdose. Assistance from friendly foreign countries / outside agencies would also need be taken into consideration.

(g) ***Early De-requisitioning***: Armed Forces should be de-requisitioned as soon as the situation in the incident/disaster-affected area has been brought under the control of the civil administration.

## Phases of Consequence Management

The channel for requisitioning of aid by civil authorities in case of WMD incidents/disaster would be the same as is being followed in the case of other kinds of aid being provided to civil authorities, during natural / manmade disasters. However, since not all units of the armed forces are trained to handle NBC related incidents, there will be a need to co-opt the specialised Response Assets which have been raised and trained for dealing with such incidents.

On receipt of information to the likelihood of a WMD strike by terrorists or if a device has already gone off, the Armed Forces will earmark a Joint Task Force – Consequence Management (JTF-CM) to deal with the incident, by nominating a commander and specialised forces.

***Joint Task Force (JTF) –Consequence Management (CM) Cycle:*** The combat commander on instructions from the JTF, which plans consequence management operations deploys the forces, executes the operations, ultimately creating conditions for transition.

**Table 8.1 JTF – CM Life**

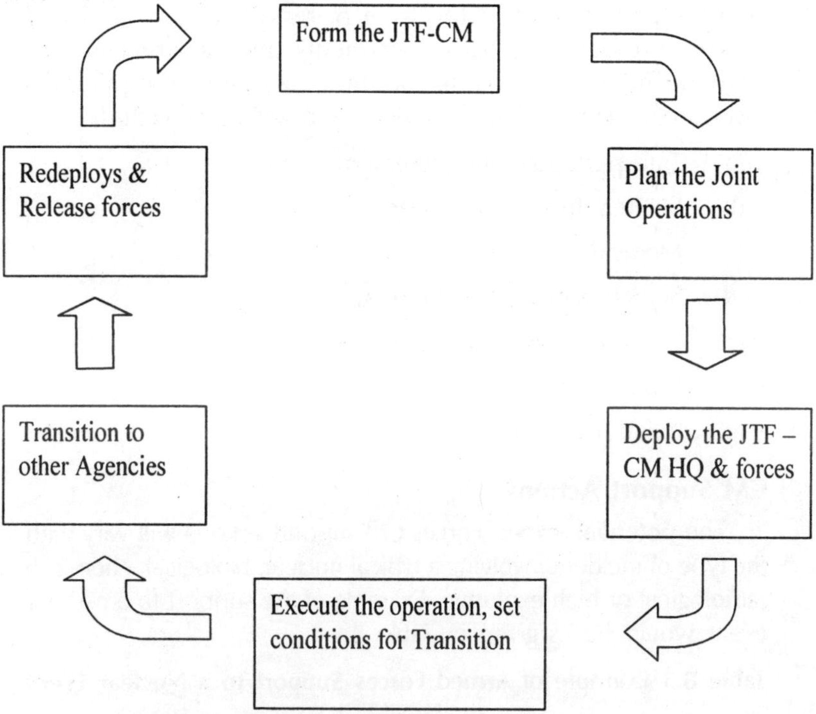


## JTF-CM Organisation

The functional organization of Joint Task Force nominated for consequence management of a domestic WMD incident will contain all elements that would be required to deal with the incident. The organisation would be as given in Table 8.2.

**Table 8.2 Notional JTF – CM Functional Organization**

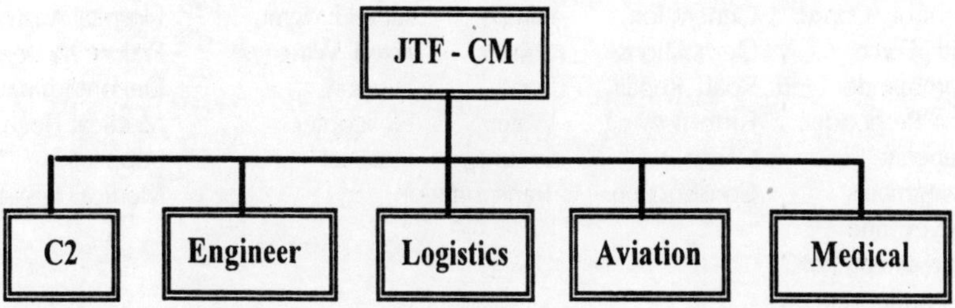

The JTF-CM support base is all-encompassing. It includes coordination support in the form of restoring command and control providing survey, assessments and reconnaissance, engineering support, transportation, aviation and medical assistance. The essential services and activities include: -

(a) Transportation and communications.

(b) Shelter, food, and water.

(c) Medical.

(d) Search and rescue support.

(e) Hazardous material response.

(f) Electrical Power.

(g) Aviation Support.

## CM Support Actions

The potential Armed Forces CM support actions will vary with the type of incident involving a typical nuclear, biological, chemical, radiological or high explosive. Example of the support to a nuclear event would be as given in Table 8.3.

**Table 8.3 Example of Armed Forces Support to a Nuclear Event**

JTF-CM

| C2 | ENG | LOG | AVN | MED |
|---|---|---|---|---|
| **Coordination Support** | **Engineer Support** | **Logistics Support** | **Air Operation** | **Medical A** |
| Command and Control, Coord. with Force, Commander and First Responder, Generate Assessments, Survey and Assessment, NBC Reconnaissance | Power Generation, Demolitions, Road Repair, Earth moving, Camp Construction. | Refugee camps, Shelter, Food, Water, Security, Transportation | Air Space Mgmt, Airfield Mgmt, Fixed Wing Support, Helicopter Support | Triage/Treatment Hospital Augmentatio Patient Movement Decontamination Medical Tech Support Medical Logistics |

## Consequence Management

The Armed Fsorces of the Union are heavily committed in fighting insurgency and terrorism since, Independence. The present geo-political situation in the country and the region points towards escalation in the terrorism ladder and may lead to use of a WMD by terrorists against the armed forces personnel or their installations. There is, therefore, a need to work out a fresh strategy to deal with such crisis. There would be a need to put less emphasis on waging conventional warfare and more on dealing with insurgencies, terrorist networks and other non-traditional threats. The proposed shift in strategic focus stems partly from a recognition of a heightened perception of such potential threats. Dealing with these types of non-traditional threats calls for specialised skills and reorganization/raising of new specialised assets. The prevailing security scenario in the region also point towards the possibilities of the armed forces and their installations coming under attack by terrorists using weapons of mass destruction (WMD) whereby they would be required to respond quickly and decisively with trained responders. Consequence management by the armed forces in response to WMD incident/disaster would involve:

(a) Protection of personnel on military installations and armed forces owned / leased facilities from WMD attacks.

(b) Respond to those attacks with trained and equipped emergency responders.

(c) Ensure that all installations are able to continue their critical operations during an attack and to resume operations after an attack.

(d) When directed / approved provide support to civil authorities.

## Anti-Terrorism (AT) Policy

The Armed Forces are the primary targets of the anti-national elements / terrorists since, most of the countries call upon them to deal with such elements as a last resort. Armed Forces in order to curb the activities of the terrorists or their supporters use their full might to finish the problem, hence are in direct conflict with them. Terrorists, whenever, opportunity presents will do their best to cause harm / damage to security forces personnel, and their assets. It becomes mandatory on the part of Armed Forces, therefore, to prepare themselves against such attacks. They need to evolve Anti-terrorism policy to deal with such types of threats.

The aim of this policy is to draw up defensive measures in order to reduce vulnerabilities and protect the Armed Forces. It involves the following: -

(a) Integrate threat / vulnerability / criticality assessment, physical protection, risk management and force employment decisions in order to detect, deter, defend against and if necessary, respond to attack by terrorists.

(b) Mandatory training and exercises for dependants, military personnel and commanders.

(c) Review and develop requirements for technology, funding, site improvement and construction enhancements.

(d) Work with C3I, in coordination with Ministry of Home, Law and External Affairs, to ensure domestic and foreign intelligence and security cooperation.

(e) Local response forces to contain terrorists attacks.

(f) Review programmes annually.

*Threat Perception*: The threat driving the AT programme is largely extracted from the past experience of attacks by the terrorists on the Armed Forces personnel and installations. The threat dimension would cover a wide spectrum to include sniper rifle, grenade / bomb blasts, Improvised Explosive Devices (IEDs), suicide attacks, attacks on facilities using trucks / boats / aircrafts as kill vehicles. The study of the terrorists' attacks will lead to an emerging pattern, the understanding of which is necessary to plan our own defensive measures.

**Threat Factors:** To realistically work out the threat levels there is a need to analyse the following threat factors:

(a) ***Operational Capability***: The operational capabilities of the terrorists need to be known in detail. These can be obtained through the intelligence sources, studying the modus operandi of terrorists, analysis of previous attacks and by monitoring transmissions of terrorists. Induction, sale / purchase of new items of armaments / equipment/ suspect material which can lead to manufacture / assembling of WMD to the terrorists will give out a general picture of their operational capabilities.

(b) ***Intentions***: The intentions of the terrorists can be gauged by their activities over a period of time by analysing the intelligence inputs and monitoring their communications and movements.

(c) ***Activity***: The increase in transmission activity and moving in of large number of suspected personnel and equipment/suspect material into a particular area and other activities will give an indication that some activity being planned by terrorists.

(d) ***Operating Environment***: The environment in which the security forces are operating will lead to decide upon the threat level in that particular area/zone. The reaction of the population of the area towards various activities of the armed forces personnel deployed in that area will enable them to gauge the level of hostility of the population towards them.

**Threat Levels:** Based on the threat factors, the threat levels could be catagorised as under:

(a) ***High***: Under this threat level, the terrorist are operationally active and they have the potential for carrying out large-scale attack including the capability to use WMD leading to mass casualties. The operating environment favours the terrorists.

(b) ***Significant***: Under this threat level there is presence of terrorist threat due to their limited operational activity and the environment is neutral.

(c) ***Moderate***: The threat is present. However there is limited operational activity. The environment favours the state.

*(d)* ***Low:*** There are no indications of presence of threat and the group activity is non-threatening. The population is cooperative and friendly towards the armed forces personnel.

The various stages of combating terrorism by the Armed Forces will include the following:

(a) ***Intelligence***: For intelligence to succeed, we must lay increased emphasis on intelligence collection, analysis and dissemination with the help of new technologies. There would be requirement to penetrate designated terrorist organisations to gain information on terrorist leadership, plans, intentions, modus operandi, finances, communications and recruitment.

(b) ***Antiterrorism***: These are defensive actions instituted at the tactical level and consist of detecting the terrorist and his network and deterring him from attacking our forces. These actions are basically defensive in nature.

(c) ***Counterterrorism***: These are offensive actions at the strategic level and include pre-emption, disrupting the terrorist plans and finally responding against him on detection with full and lethal force.

(d) ***Consequence Management***: The final action to be taken by the Armed Forces in case their installation is attacked. It would include protection of assets, mitigate the effects of terrorist attack and finally restore full scale functioning of the effected installation.

The integrated approach to deal with a terrorist incident involving WMD would be as follows: -

**Figure 8.4 Armed Forces Integrated Approach**

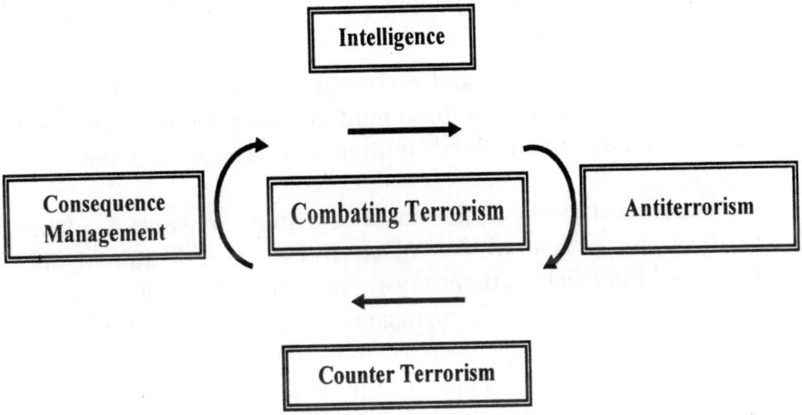


## Assets Protection Preparation

The armed forces assets are vulnerable to attacks by the terrorists. With increase in frequency of suicide attacks against these assets, there is a need to have physical security standards for protection of fixed facilities / installations. These standards are then required to be scrutinised by an independent agency, which should have sufficient resources to assist the field commanders with assets/installation protection measures. The role of the agency would be to study the weaknesses of an installation and recommend suitable measures to further enhance the protection of the facility against any terrorist attacks. The comments/remarks of the agency would in a way reflect on the performance of the personnel responsible for the security of the installation. This agency should have ladequate funds to manage research and development to enhance physical protection and physical security measures.

The agency will be responsible for proceeding integrated expertise as a catalyst to effect changes in the anti-terrorism policy by:

(a) Assessing installations to determine vulnerabilities and provide executable options.

(b) Provide technical assistance throughout to fill any gaps in policy, doctrine, training and education and technology development.

(c) Provide training and technical assistance to combatants / services for establishment of assessment teams and increase the general anti-terrorism knowledge base.

The armed forces of the nation are strategic assets available to the state to deal with WMD terrorism. A well trained and equipped force acts as deterrence. They are effective instruments to carry out consequence management in case the terrorists are able to strike with weapons of mass destruction. The armed forces assets are vulnerable to terrorist strikes and hence, they must have an effective anti terrorism policy to ensure that their functioning is not affected by their actions. With the increase in suicide terrorism, the armed forces installations require an independent agency to critically asses their shortcomings and recommend and initiate suitable measures so that their safety and functioning is not jeo parachised or affected even under terrorist attack. Fool proof safety measures and security drills against such attacks are themselves a dependable deterrence.

# 9

# Medical Management of WMD

*Humans, regrettably, have used available technologies for destructive as well as beneficial purposes throughout history.*

**Journal of the American Medical Association, 1997**

Nuclear, biological and chemical weapons are characterised as weapons of mass destruction. These weapons are characterised by the occurrence of mass casualties overwhelming the capabilities of the existing medical care infrastructure. Rapid industrialisation and urban proliferation has resulted in hazardous chemical industries being located in even thickly populated localities. Accidents in these hazardous chemical industries can affect the population located in their near proximity and under favourable meteorological conditions even in far-flung areas - the Bhopal Gas Tragedy of 1984 being a very prominent example of such tragic accidents.

Effects of use of mass destruction weapons would be the worst of all manmade disasters. Unlike the conventional explosives, these will be thousand times more devastating. Usage of protective measures thus, will be required for a considerable period, due to long term effects of these weapons on personnel and denial of resources, if contaminated. Use of Weapons of Mass Destruction (WMD) results in substantial increase in the casualties owing to: -

(a) Destructive nature of the weapon.

(b) Contamination of Conventional Casualties.

(c) Psychiatric casualties.

Medical management in NBC Warfare, however, poses a number of challenges – the number of casualties occurring practically instantaneously will be too large, the remaining and available medical resources grossly insufficient, the lack of experience in dealing with such types of situation and problems of the combined injury syndrome i.e. contamination of conventional casualties. Mass casualty medicine using simplified and standardised regimens will have to be evolved to accomplish what is now done by labour intensive and resource demanding means.

## Nature of Casualties

Nuclear weapons usage results in casualties due to thermal effects, blast wave and nuclear radiation to the tune of 50, 35 and 15% respectively. One important difference between a nuclear and conventional (chemical) explosion is the appearance of an appreciable portion of the energy as thermal radiation in nuclear explosion. Consequently, the temperature reach, in case of nuclear explosion, is much higher and may reach millions of degrees centigrade. The blast waves can cause injuries due to direct blast wind effects like flying debris, crush and transactional injuries. The thermal injuries can result in flame burns and eye injuries like flash blindness and retinal burns. The nuclear radiations can cause a combination of syndromes clubbed under the nomenclature of Acute Radiation Syndrome. The treatment modalities of the effects due to exposure to nuclear weapons will follow the conventional methods except in case of injuries with concomitant contamination, calling for quick decontamination procedures. The other aspects of nuclear weapons casualties like the large number of burns cases and the delayed cellular effects of radiation necessitate the establishment of specialised medical departments like bone marrow centres, malignant disease treatment centres, haematology centre, trauma centre etc.

There is nothing mysterious about the medical problems that would be created by a successful biological weapons attack. The situation would require medical management on a mass basis rather than the administration of individual care but the basic principles of medical therapy would apply. The major problems of medical care following an effective biological attack will be those produced by the extremely large number of casualties by clandestine operations such as introduction of organisms into the air handling systems of a building or contamination of food or water supplies, are possible and could have a distinct harassing effect. Public health measures and general preventive medicine principles would be of some importance but no special measures are advocated for biological weapon casualties' management.

The nature of casualties exposed to chemical warfare agents, by and large depend upon the type of agent to which exposed. The nomenclature of the agents gives an indication of the system being targeted i.e. - nerve agents, blister agents, blood agents, chocking agents etc. Treatment will follow the conventional methods. However, while most of the known chemical agents have specific antidotes, quite a few of them are not in common use. Hence, planning for medical management in the chemical agent scenario has to cater for stocking and provisioning of specific antidotes, that might be needed to meet such emergencies.

Careful advance planning is essential for a country to be adequately equipped to manage the threat of the consequences of deliberate release of biological or chemical agents. A central consideration in such preparedness planning is that, it is neither possible nor necessary to prepare specifically for attack by all possible biological and chemical agents. If, a country is seeking to increase its preparedness to counter the effects of biological and chemical attacks, the targeting of its preparation and training on a limited but well chosen group of agents will provide the necessary capability to deal with a far wider range of possibilities.

## Consequences of Using Biological or Chemical Agents

***Short-Term Consequences*:** The most prominent short-term effect of biological or chemical agents is the large number of casualties that they can cause, and it is this characteristic that determines most preparedness strategies. The potential for overwhelming medical resources and infrastructure is magnified by the fact that the psychological reaction, including possible terror and panic, of a civilian population to biological or chemical attack may be more serious than that caused by attack with conventional weapons. Psychological support strategies combined with risk communication are an integral part of the services needed to manage the many exposed and non-exposed casualties who may have to be treated or tackled.

***Long-Term Consequences***: Some biological and chemical agents have the potential to cause physical or mental illness that either remain, or only become evident, months or years after the weapons have been used. They may enter the potential for harm of biological or chemical weapons beyond their immediate target both in time and space. The long-term health consequences of release of biological or chemical agents may include chronic illness, delayed effects, new infection diseases becoming endemic and effects mediated by ecological changes. The delayed effects in persons exposed to certain biological and chemical agents, depending on the dose received, may include carcinogenesis, teratogenesis and perhaps mutagenesis. If, biological agents are used to cause disease that are not endemic in the country attacked, this may result in the disease becoming endemic, either in human population or in suitable vector such as arthropoda and other non-human hosts such as rodents, birds or cattle.

There is a lack of scientific awareness, preparedness and funding. Stocks of antibiotics and vaccines against known pathogens that are essentially needed are inadequate to meet with ever the periodical natural high incidence, let alone in event of a bio terrorist attack. The situation is worse in the developing countries. Throughout most of the world, the public health infrastructure is stretched to its limits coping with natural health hazards. In 1998, a quarter of the world's 54 million deaths were due to infectious diseases and in developing countries such disease caused one in two deaths. Against a such a backdrop, the additional threat to public health against a disease caused in a country by biological warfare might be no more than a slight addition to the existing burden. However, it might also be on such a scale or of such a nature as to be beyond the capability of the health care system to cope lap with. For deliberate release or threats of release of biological agents, a spectrum of threat can therefore be envisaged that ranges between these two extremes, relative insignificance at one end, mass destruction of life or mass casualties at the other. Where along this spectrum a particular biological menace is situated will be determined by the characteristics of the agent and the way it is used, and by the vulnerability of the threatened population, reflecting such factors as the health status and degree of preparedness of that population. Particularly threatening would be the possibility of a pandemic resulting from the intentional or inadvertent release of infective agents that cause contagious diseases, such as smallpox, for which effective hygienic measures, prophylaxis or therapy may be unavailable. Towards the mass destruction end of the spectrum, remedies or counter-measures may be beyond the resources of many countries.

## Consequences of Using Chemical Weapons.

The most prominent short-term effect of chemical weapons is the large number of casualties that they can cause and magnification of the potential for overwhelming medical resources and infrastructure. An illustrative example of nature of the short term consequences of urban attack with chemical agents is the 1995 episode of Tokyo Subway incident where sarin was used by Aum Shinrikyo which lead to 12 mortalities and caused more than 5000 people to seek medical care. The majority of the 5000 people seeking help, many of them with psychogenic symptoms, were worried that they might have been exposed.

The long-term health consequences of release of chemical agents may include chronic illness and delayed effects. The potential for chronic illness after exposure to some toxic chemicals is well known. Follow-up of Iranian victims of Iraq–Iran War has revealed debilitating long-term disease of the lungs (e.g. chronic bronchitis, asthmatic bronchitis, pulmonary fibrosis etc), eyes (delayed mustard gas kiratitis with blindness) and skin (dry and itchy skin, pigmentation disorders etc). Deaths from pulmonary complications were still occurring as late as 12 years after all exposure had ended. The delayed effects in persons exposed to certain chemical agents may include carcinogenesis, teratogenesis and perhaps mutagenesis.

## Consequences of Using Biological Weapons

The most prominent short-term effect of biological weapons use is the large number of casualties that they can cause, and it is this characteristic that determines most preparedness strategies. The potential for overwhelming medical resources and infrastructure is magnified by the fact that the psychological reaction, including possible terror and panic incivilian population to a biological attack may be more serious than that caused by attack with conventional weapons. The possible long-term consequences of the use of biological weapons including delayed, prolonged and environmentally mediated health effects long after the time and place that the weapons were used are more uncertain. The long-term health consequences of release of biological agents may include chronic illness delayed effects, new infectious diseases becoming endemic and effects mediated by ecological changes. If, biological agents are used to cause disease that are not endemic in the country attacked, this may result in the disease becoming endemic, either in human population or in suitable vectors such as arthropoda and other non-human hosts such as rodents, birds or cattle. Finally, there is the possibility of effects mediated by ecological changes. New foci of disease might become established as a result of ecological changes caused by the use of biological agents infective for humans and animals, or as a result of the use of anti-plant agents by means of reduction in quality and quantity of food supply or economic effects through direct effects on agriculture or through indirect effect on trade and tourism.

## Public Health Preparedness And Response

The initial response to a deliberate release of infections or toxic agents targeted against civilian population is largely a local responsibility in many parts of the world. Local authorities are in the best position to deal with such events, and will generally be held accountable should the incident be mishandled. It is the responsibility of local officials to ensure that response systems and plans are in place before an incident actually occurs. In contrast to chemical agent attack where the first to responders are likely to be police, fire departments and emergency medical personnel on or near the scene, the first to respond to an initially undetected attack with an infective or toxic agent having only delayed effects are more likely to be regular health-care providers, including nurses, doctors physician and hospital accident and emergency personnel who may be located in widely separated places. A covert release of a biological agent will be more likely to become apparent over a longer period of time e.g. days or even weeks, and will probably take the form of the appearance of cases of infectious disease. Because some victims are likely to move around during the symptom-free incubation period after exposure, cases of the distant ones, and the full picture may become evident only after information, medical reports and surveillance data from many areas, have been combined collated and analysed. While chemical weapons can place a great burden on public safety personnel and biological weapons on the public health infrastructure, they can both place an extraordinary burden on the local health care delivery system. Because victims of a chemical attack may be affected immediately, a rapid response will be required, in which the main emphasis will be on evacuation, contamination control and early medical treatment. The ability to respond to biological incidents depends on preparedness (what needs to be considered long before an incident takes place) and response (what needs to happen after a warning of a pending release is received, or after the release has actually occurred).

## Preparedness

***Threat Analysis***: Threat analysis is a multidisciplinary activity with inputs from the country's law enforcement, intelligence and medical and scientific communities and aimed at identifying those who may wish to use biological agents against the population, the agents that may be used and the circumstances under which they may be used. This requires active liaison between law enforcement, security and health agencies with the local authorities. The level of threat that exists in also a function of the potential vulnerability of the community concerned. Vulnerability analysis will identify potential scenarios as well as weaknesses in the system that may be exposed to biological hazards and determine the current ability to respond to and manage the emergency.

***Pre-emption of Attack***: The establishment of a biological and chemical response system is in itself a pre-emptive risk reduction strategy. The mere existence of effective ability to respond to and manage an incident reduces the risk of biological attack. Intelligence on terrorism relies heavily on human sources, as terrorist activities may be much less conspicuous and therefore, more difficult to detect by reconnaissance satellites etc. An important pre-requisite for pre-emption is the existence of national legislations that renders the development, production, possession, transfer or use of biological weapons a crime and that empower law-enforcement agencies to act where such activities are suspected before an event actually occurs.

***Preparing To Respond***: Pre-empting efforts not withstanding, a preparedness programme may be necessary which will require the acquisition of equipment and supplies, the development of appropriate procedures and training. Communities will need to examine their existing public health plans and current training for adapting them in the light of features unique to deliberately released biological agents. Most civilian health care providers may have little or no experience of biological weapons and may, therefore, not suspect especially in the early phases of an incident, that a patient's symptoms could be due to such weapons. Tshere is, therefore a need to train health care workers in the recognition and initial management of biological casualties and for a rapid communication system that allows for sharing of information immediately when an unusual incident is suspected. Education and training must cover the general characteristics of biological agents, the clinical presentation, diagnosis, prophylaxis and treatment of diseases that may be caused by deliberate agent release and sample handling decontamination and barrier nursing. Training, planning and drills should be directed at physicians and staff for the management of mass casualties. Large scale distribution of medication and supporting the local authorities in vaccination programmes, providing the necessary education and training is expensive and manpower-intensive, yet may be the most cost-effective method of medical preparation for biological attacks. Such training will also be the co-option of an approach to prevent anxiety and fear in health-care workers, something that might be expected after a bio-terrorist incident and that could disrupt the provisioning of health care services.

***Public Awareness***: If it is to have any chance of success, a plan for providing information to the public and thus, demystifying the subject of biological weapons needs to be drawn up well before an incident occurs. If this is to be effective, the public needs to know how they are expected to act if, an attack takes place. The communication plan may include radio and television broadcasts or the distribution of folders brochures to the public describing the potential threat in plain, unemotional language, clear advice should be given on how the alarm will be raised and what to do if that happens. These must contain explicit and exhaustive instructions on channels of communication and clearance procedures for potentially sensitive information.

A question that is raised often is "whether it is wise to make the biodefence strategy and state of preparedness a public knowledge". It is felt that the bioterrorist also gets to know of the detailed plans and built up defences and would act by-passing them. There are atleast three considerations in favour of transparency:

(a) Firstly, biodefence is aimed at facing a bioterrorist attack, but preventing this is the better choice. Well-advertised preparedness against the best arsenal of a bioterrorist would be a deterrent. We are fighting bioterrorism and not the bioterrorist.

(b) Secondly, if the public comes to know of what the government and other agencies have been doing for their safety, their confidence in our public institutions, which is a very important psychological factor, would grow.

(c) Thirdly, awareness of the governmental efforts may inspire private organisations and individuals to add their own bit to the national effort.

***Validation Of Response Capabilities***: As with preparation for any high consequence but low frequency incident, it is a major challenge to prove and validate response capabilities if they are not being constantly practiced or used. In addition, careful analysis of actual incidents, wherever they occur, should provide valuable information and the lessons learnt should be incorporated into our future planning. The danger of making the response to biological incidents the task solely of dedicated specialized response units is that the relative infrequency of callout could lead to deterioration of skills and more seriously, excessive centralization may risk increasing the time taken to react. A readiness and response strategy should aim at enabling the local public health, emergency response and other authorities (firefighting, police and ambulance services and civil defence) to respond to and manage the incident scene in its early phases, specialised functions being performed later by a dedicated mobile biological unit.

## Response Before any Overt Release of a Biological or Chemical Agent

If warning of an impending release of biological or chemical agent is received, a number of activities can and should be carried out before the release, if any, actually happens. The sequence in which these activities are performed will depend on the particular circumstances of the incident. The first indication of an incident may be a warning, or the finding of an unusable device or unusable materials as results of normal activities within the community such as the response to a fire or the discovery of a strange package. One or more of the following may be required.

(a) ***Analysis of The Available Information***: All the information available needs to be assessed by an appropriate group including the police, the intelligence services and technical experts who should have been trained to work together to analyse such information by means of realistic and credible exercise. Such a small group of analysts and experts will be able to evaluate the threat or the information on the incident and advice on appropriate action and the mobilisation of specialist assistance, and may also help to avoid inappropriate responses to hoaxes.

(b) ***Initiation of Search Procedure***: If sufficient information was given in the warning and the analysis warrants such action, it may be appropriate to search for a suspect device at a particular location.

It may also be appropriate to search for those responsible for the warning or witness who may have seen them.

(c) ***Establishment of Cordon***: Again depending on the circumstances and the information available, it may be appropriate to evacuate people from the area at risk and to establish an exclusion zone.

(d) ***Early Identification of the Nature of the Hazard***: If a device or unusual package is found, it will be important to decide as soon as possible whether the impending hazard is chemical or biological in character (or even a mixture of the two). The presence of explosive, either as the primary hazard or as the disseminating charger for a toxic/infective agent, must also be considered, together with the possibility of the device containing a radioactive hazard. The appropriate specialists can then be called in to help in managing the incident, and the appropriate protective equipment selected. For example, an oronasal mask provides adequate protection against a particulate biological hazard while a respirator and full protective clothing may be required to protect against a persistent chemical agent.

(e) ***Risk Reduction and /or Neutralisation***: Depending on the nature of the device, the possibility of reducing the risk, or neutralising the potential hazard, through containment or other mitigation and neutralisation approaches should be considered. Whether it should be managed on-site or moved to a specialised facility would be a decision for specialists (equipment is available that allows on-site controlled and contained detonation of devices, together with decontamination of toxic/infective contents). Wherever possible, sampling for analytical and forensic purposes should be accomplished before destructive neutralisation.

## Distinguishing Features of Biological and Chemical Incidents

In the earliest phases of a release (and particularly when it is covert), it may be difficult to distinguish between a biological and a chemical attack. As a general rule, chemical attacks are more likely to produce simultaneous and similar symptoms in a relatively restricted area near the point of release relatively soon after release. Biological attacks are more likely to result in the appearance of ill individuals at medical centres and/or doctors' over a longer period of time and a much larger area. Symptoms resulting from exposure to chemicals with delayed effects will obviously be much more difficult to distinguish from those of an infectious diseases. While there are no definitive and invariable distinguishing features, the indicators shown in Table 9.1 may help in deciding whether a biological or chemical attack has taken place.

**Table 9.1 Differentiation of Biological and Chemical Attack**

| Indicator | Chemical Attack | Biological Attack |
|---|---|---|
| Epidemiological Feature | Unusual number of patients with very similar symptoms seeking care virtually simultaneously (especially with respiratory ocular, cutaneous or neurological symptoms, e.g. nausea, headache, eye pain or irritation, disorientation, difficulty with breathing, convulsions and even sudden death) | Rapidly increasing diseases incidence (over hours or days) in a normal population |
| | | Unusual increase in people seeking care, especially with fever, respiratory, or gastrointestinal complaints |
| | Clusters of patients arriving from a single locality | Endemic diseases rapidly emerging at an unusual time or in an unusual pattern |
| | Definite pattern of symptoms clearly evident | Patients· with a relatively uncommon diseases that has bioterrorism potential (patricularly those listed in Annex 3) |
| Animal indicators | Sick or dying animals | Sick or dying animals |
| Devices, unusual liquid spray or vapour | Suspicious devices or packages Droplets, oily film unexplained odour Low clouds or fog unrelated to weather | Suspicious devices or packages |

## Response to Biological or Chemical Attack Incident

***Hazard Identification***: Detection and identification are necessary to determine the nature of the chemical or biological hazard being confronted. Detection strategies may include the use of a variety of devices that can provide an early indication of the agent involved. This is needed to guide operational response activities. A large variety of devices are available, ranging from simple colour-changing paper to sophisticated electronic contamination monitors. Detection strategies must be linked to warning or alert mechanism that will be used to activate response. Whether by primary responders, specialist responders or the population. In case of biological agents, prompt identification of the agent involved is required to ensure that the appropriate preventive and medical measures are taken. Because early diagnosis of biological exposure will be important in the choice of treatment and response, establishment of a reference laboratory (or a network of laboratories in large areas) in which potential agents can be identified should be catered for in the preparation phase, the extent to which laboratory support will be able to and initial diagnosis and treatment will depend on both the level of pre-incident preparation and availability of a network of diagnostic laboratories.

***Evaluation of Potential Spread***: In an overt chemical release, an important component of exposure assessment is the prediction of the spread of the agent cloud. A variety of computerised prediction models are available which can be used to indicate where effects are likely to be greatest and to direct the deployment of resources. If the incident involves the release of a biological aerosol, computer modeling may help to predict the spread of the aerosol particles. The first steps must, however, be to gather information on the wind direction and speed and on possible sources of the aerosol. Retrospective analysis may indicate that cases originate from specific areas and may be a valuable indicator of an up-wind site of original release. In case of agent that has potential for person-to-person transmission, standard epidemiological methods should then be used to predict the probable spread of the disease, and medical recourses mobilised and deployed accordingly.

***Protection of Responders***: Individual protective equipment (IPE) must be available to responders and must allow them to carry out a wide range of activities in a contaminated area without becoming casualties. In areas where the threat is significant, it may be necessary to make collective protection facilities available i.e. large protected air where people can shelter without the need of an IPE. In addition to compromising the ability to manage the incident, the occurrence of infection in health care workers may lead to the perception among the population that health centres and hospitals themselves constitute a high-risk source of infection. This may discourage potentially infected person from seeking treatment from the local health care providers, and then to travel to other health care facilities, thereby increasing the risk of secondary transmission if the infection is contagious. During the spread of a biological aerosol, the primary route of exposure will be via the airways and respiratory tract. Particulate filters are generally adequate for respiratory protection against biological agents. Universal precautions for dealing with potentially infective materials should be taken and protection of responders should be based on the standard principles of barrier nursing and infection control. Vaccination or prophylactic antibiotic treatment of these involved in response may have to be considered. Pre-attack vaccination of health care providers may be considered if, appropriate vaccines are widely available.

***Infection/Contamination Control***: The most distinctive element of disaster management for chemical incidents is contamination control, which requires:

(a) The rapid establishment of a well demarcated "hot zone" (with clearly visible "clean" and "dirty" areas).

(b) The limitation of contamination spread by means of strictly controlled entry and exist procedures.

On-site decontamination procedures ensures that all persons or items leaving the dirty areas are cleaned and monitored before entering the clean environment. If agents of transmissible (contagious) diseases are released, basic hygiene and infection control measures, e.g. washing hands after contact, avoiding direct contact with secretions from infected individuals, keeping exposed persons away from public places, and isolating suspected or symptomatic cases, may by essential in limiting secondary spread. The dissemination of such basic information on the precautions is necessary, not only to health-care providers but also to the general public, which will be an important step in infection control. The population should be told what signs and symptoms to watch out for and who to call or where to go if they appear. Lack of specificity in such advice to the public may result in local health facilities becoming overwhelmed by uninfected patients.

Large-scale evacuation as a preventive measure is not likely to form part of the response to biological incidents. Where contagious disease is involved, it may aggravate the situation by increasing both the spread of infection and the number of secondary outbreaks. Movements of patients should be restricted to the minimum necessary to provide treatment and care.

Special measures may be required to limit the nosocomial spread of such disease as the viral haemorrhagic fevers (e.g. Ebola or Marburg), plague and small pox. The frequent suggestion that special rooms under negative pressure should be provided is impractical because of the sheer number of probable cases. Provision may be made to care for patients at sites other than health-care centres, such as gymnasia, sports arenas or at home.

Immediate decontamination for people who may be exposed to biological attack is not so critical as it is for chemical casualties, since, biological agents are non-volatile, are difficult to re-aerosolise and leave little residue on skin or surfaces. Many pathogens deposited on surfaces will rapidly die, though some may survive for longer periods. However, it would be prudent to be prepared to decontaminate both materials and persons, particularly if a site of release can be identified. Defining a "hot zone" (as in hazardous-materials incidents) may be extremely difficult or impossible, and it may not be possible to define the contaminated zone until the outbreak has been characterised. At or near the release point of a biological agent, where large particles may have been deposited, area decontamination (or whole-body decontamination of persons who were present in the area) may be appropriate. Decontamination solutions used for chemical decontamination will usually also be suitable for biological decontamination. Hypochlorite is the recommended disinfectant for use in outbreak response. An all-purpose disinfectant should have a concentration of 0.05% (i.e. 1 g/litre) of available chlorine, a stronger solution with a concentration of 0.5% (i.e. 10 g / litre) available chlorine being used for example, in suspected outbreaks of Lass and Ebola virus diseases. The use of the solution with 0.5% available chlorine is recommended for disinfecting excreta, cadavers, and spills of blood and body fluids, and that of the solution with 0.05% available chlorine for disinfecting gloved or bare hands and skin, floors, clothing, equipment and bedding. Most experts now agree that water, or soap and water, may be adequate, and probably safer, for the removal of most biological agents from human skin. Buildings can be decontaminated by means of chlorine-based liquid sprays, formaldehyde vapour produced by heating Paraformaldehyde, or other disinfecting fumigants. Because of the lack of other effective tools, the decontamination of a building may be psychologically beneficial. It may, however, be extremely difficult to certify that a building is clean after an agent release. In addition to the standard principles of barrier nursing referred to above for highly transmissible agents, the disposal of waste materials, safe burial practices, and cleaning or disinfections of patients' clothing should be considered.

Where transmissible-disease agents are involved, quarantine of the affected area via the establishment of a sanitary cordon may need to be considered. The coordinated efforts of several public service groups will be required to inform the people affected, control water and food supplies, regulate the movement of people into and out of the area, and establish medical services. In addition, where there is a danger of the international spread of human diseases, the provisions of the International Health Regulations (IHR), currently under revision, should be borne in mind. The IHR provide an essential global regulatory framework to prevent the international spread of diseases through permanent preventive measures for travellers and cargo, and at border crossing points.

## Triage

Any suspected or actual dissemination of biological agents is likely to lead large number of people seeking care. The development of scientifically sound case definition(s) suitable for the local circumstances and the definition of the population at risk of becoming ill are very important for triage (the initial reception, assessment and prioritisation of casualties). Such information can generally be gathered from the epidemiological description of the outbreak, or sometimes from more specific surveys. Fear and panic can be expected in genuinely symptomatic patients, the public and the health care providers involved. All health-care facilities will need to plan in advance for dealing with overwhelming numbers of people seeking care or advice simultaneously, and to ensure that resources are used to help those who are most likely to benefit. Both psychological support and active treatment of anxiety will play an important part in the triage process.

## Medical Care

Medical care includes prophylaxis (pre-exposure treatment for high risk personnel to prevent or minimise the effects of exposure). There are not many examples of true prophylaxis, but certain medications (e.g. Pyridostigmine Bromide) can improve the response to treatment of those affected by nerve agents. Initial pre-hospital treatment will provide symptomatic and life saving support to allow decontamination and transport to medical centre. If the nature of the substance is known, specific treatment protocols may be required for on-site emergency antidote administration (possibly using auto injector) and definitive treatment of the medium and long-term effect of exposure.

The specific medical treatment of exposed individuals will depend entirely on the nature of organism involved. Immunisation or prophylactic antibiotic treatments of certain segments of the population (contacts, health care personnel and first responders) against potential biological agents may be warranted. This treatment will depend on the availability of such treatment and its effectiveness against the agent involved, e.g. immunisation will be an important means of controlling an outbreak of smallpoxor plague, and all those who enter hospitals where patients are housed and treated should be immunised against these diseases.

Because immunity generally takes several weeks to develop fully after vaccination, drugs (antibiotics) and symptomatic care may be the mainstay of management. Immune serum may be used to confer passive immunity.

If, stockpiles of antibiotics or vaccines have been prepared or identified, plans for their distribution must be activated. In essence, the choice is either to take the drug to the potentially exposed person or for the person to come to the drug. The latter option generally requires fewer personnel. The stocks should be larger than needed to treat only those exposed because it may be difficult to distinguish between those who have actually been exposed and those who simply believe themselves to have been exposed. Cases may be much greater in number than the total number of available hospital beds and additional care facilities may need to be established.

## International Assistance

The management of a large-scale outbreak, whether of natural, accidental or intentional origin, will be beyond the resources of many countries. An early decision to enlist the assistance of international aid may save many lives. World Health Organisation (WHO) does offer public health assistance to countries experiencing outbreaks of infectious diseases , and such aid will be available regardless of the source of the outbreak.

## Medical Management of WMD Casualties

The principles of medical management of mass casualties are applicable in a WMD scenario but with some significant changes with reference to aspects of triage, decontamination of casualties and disposal of contaminated waste and dead bodies. Qualitatively, the difference between NBC and conventional weapons is of even greater significance than the quantitative differences. Apart from the direct effects of NBC weapons, the fallout can be transported to greater distances from the actual site. Moreover, creation of contaminated environment will be an obstacle to rescue operations and effective care of injured survivors and will have harmful or lethal effects long after the use of weapons.

## Triage

In a mass casualty situation, sorting of the casualties and laying down priorities for their treatment is important. Triage is practiced only when a mass casualty situation occurs and the needs of the casualties for care overwhelm the medical capabilities to provide that care. It ensures the greatest care for the greatest number and the maximal utilisation of medical assets, personel, supplies and facilities. In a conventional warfare scenario, what is being taught and being followed, is for prioritising the casualties into three categories namely I, II and III. However, the magnitude of the casualties in an NBC scenario and the atypical effects of NBC weapons warrant a change in the triage philosophy. Thus, in an NBC scenario, the casualties need to be prioritised into:

(a) ***NBC Priority I (IMMEDIATE)***: Casualties who require life saving care within a short time, when that care is available and of short duration.

(b) ***NBC Priority II (DELAYED):*** Casualties who are in need of major or prolonged surgery or other care and who will require hospitalisation, but delay of this care will not adversely affect the outcome of the injury.

(c) ***NBC Priority III (MINIMAL)***: Casualties who have minor injuries, can be helped by paramedical staff, will not be evacuated and will be able to return to duty shortly.

(d) ***NBC Priority IV (EXPECTANT):*** Casualties with severe life threatening injuries who would not survive despite optimal medical care. The above priorities being common to NBC weapons it is important to delineate the chemical agent casualties into priorities, keeping in mind the toxic effects of the different groups of these agents as the severity of their effects are quite different. To simplify, the prioritisation of chemical casualties depending on the agent to which exposed could be as follows: -

### (i) IMMEDIATE (PRIORITY- 1)

(aa) ***Nerve Agent***: Casualties talking but not walking (severe distress with dyspnoea, twitching and/or nausea and vomiting): moderate to severe effects in two or more systems (e.g. respiratory, muscular, G 1): circulation intact. Not talking (Unconscious), not walking, circulation intact. Not talking, not walking: circulation not intact (If treatment facilities are available; if not classify as expectant}:

(ab) ***Cyanide***: Severe distress (unconscious, convulsing or post ictal, with or without apnoea) with circulation intact.

(ac) ***Vesicant***: Airway injury: classify as immediate help can be obtained.

(ad) ***Phosgene:*** Classify as immediate if help can be obtained.

### (ii) DELAYED (PRIORITY -II).

(aa) ***Nerve Agent***: Recovering from severe exposure antidotes or both.

(ab) ***Vesicant***: Skin injury greater than 5% but less 50% (liquid exposure) of body surface: any body surface area but (vapour exposure); most eye injuries, airway problem starting greater than 6 hours after exposure.

(ac) ***Cyanide***: Recovering: has survived more than 15 minutes after vapour exposure.

### (iii) MINIMAL (PRIORITY - III).

(aa) ***Nerve Agent***: Casualty talking and walking: capable of self- aid: return to duty imminent.

(ab) ***Vesicant.*** Skin injury less than 5% of body surface area in non-critical areas: minor eye injuries, in or upper airway injury.

### (iv) EXPECTANT (PRIORITY –IV)

(aa) ***Nerve Agent***: Not talking, circulation failed with adequate treatment resources, should classify as immediate.

(ab) ***Vesicant***: Over 50% body surface area skin injury from liquid: moderate to severe airway injury particularly with early onset i.e., 6 hours after exposure.

(ac) *Cyanide*: Circulation failed.

(ad) ***Phosgene***: Moderate to severe injury with early onset.

Triage of casualties of NBC agent use is based on the same principles as the triage of conventional casualties. However, it must be understood that triage refers to priority for medical or surgical care, not priority for decontamination. All NBC casualties and chemical casualties in particular require decontamination.

## Decontamination

A major problem in the management of NBC casualties is tackling of the contamination hazards by the NBC agent used. Decontamination, is thus, essential both to terminate this hazard to the individual as also to prevent the reach of this contamination. The aim of decontamination is to rapidly and effectively render harmless or remove toxic substance both on person and equipment. Decontamination is defined as the reduction or removal of radiological dust, chemical or biological agents so that they are no longer medical hazards. Agents may be removed by physical means or be neutralised chemically. Decontamination of the skin is the primary concern, but decontamination of eye and hands must also be carried out when necessary. Decontamination procedures for the various contaminants are as follows: -

(a) ***Radiological Contamination***:

Radiologically ontaminated patients are those whose outer body surfaces have been contaminated with fallout from a nuclear detonation. These fallout residues, in the form of dust, ashes, dirt or mud, will be loosely adherent to the clothing and skin of the patient. Once these residues have been removed, the patient does not present a radiation hazard. The practical decontamination of radiologically contaminated patients is easily accomplished without interfering with the required medical care. 90 to 95% of the decontamination can be accomplished by simply removing the outer clothing and protective equipment. The contaminated clothing once removed, can be placed in bags, tagged and removed to a remote section. The second phase of decontamination consists of washing or wiping the patient's face and hands with water and soap. This should have the patient 98% decontaminated. The third phase of decontamination consists of washing the hair or clipping the hair and washing the scalp. This would be required if the patient arrives without headgear.

**(b) Procedure for Skin Decontamination.**

- Cut open the clothing of the injured/contaminated person.
- Put all the clothing/swabs in a plastic bag. Wash the open wounds if any with saline. Irrigate with water.
- Ensure that the decontamination is proper with the help of radiation detection equipment.
- Clean the skin and decontaminate by progressive cleansing agents namely mild soap or scrub soap, EDT A and sodium bisulphite.
- Gargle with 3% $H_2O_2$ for pharyngeal contamination. Rinse eyes and nose with saline.
- Caution should be taken not to disrupt the integrity of the skin while scrubbing because disruption can lead to incorporation of the radioisotopes into deeper layers of the skin. The prevention of incorporation is of paramount importance. The inhalation or ingestion of radioactive particles is a much more difficult problem, and resources to deal with it may not be available in a field situation. Decorporation is the procedure by which ingested radioactive particles and isotopes are prevented from entering their target organ and subsequently, expelling them from the body.

**(c) *Decorporating Agents*:**

The following decorporating agents are recommended for tackling internal emitters entered in the body during the nuclear explosion:

- Barium Sulphate -for strontium and radium in a dose of 300 gm orally.
- Frusemide- for sodium and potassium in a dose or 40 gm a day orally.
- Lugols Iodine -for iodine in a dose of 1 ml a day orally for 7 days.
- Ammonium Chloride -for strontium in a dose of 2 gm three times a day for 6 days.
- Dimercaprol -for Mercury, Gold, Arsenic, Bismuth, Lead, Chromium, Polonium and Nickel in dose of 100 mg 4 hourly for 2 days IM, followed by 12 hourly for 1 day and then once daily for 5-10 days.
- Pencillamine -for Copper, Iron, Mercury, Lead, Gold in a dose of 250 mg orally 6 hourly for 8-10 days.
- Sodium Bicarbonate for Uranium -3.5 gm in 250 ml.
- Desferrioxamine- for manganese, Iron, Cobalt at the dose of 1 gm IV Saline.
- Prussian Blue for Calcium and thallium in a dose of 1 gm TDS up to maximum of 10 gm a day.
- DTPA in a dose of 250 mg IV daily for manganese, cobalt zirconium.

## Biological Agent Contamination

Decontamination of personnel after a biological warfare attack is a lesser concern than after a chemical agent attack because most biological warfare agents are not dermally active (the tricothenes mycotoxin being an exception). Still decontamination remains an effective way to decrease the spread of infection from potential secondary aerosolisation. Biological warfare agents can be decontaminated by chemical and physical methods.

(a) ***Chemical Method***: Chemical decontamination renders biological warfare agents harmless by use of disinfectants. Dermal exposure to a suspected BW agent should be immediately treated by soap and water for decontamination. It is important to use a brush to ensure mechanical loosening from the skin surface and then to rinse with copious amounts of water. This method is often sufficient to prevent contact infection. The contaminated areas should then be washed with a 05% hypochlorite solution with a contact time of about 10-15 minutes. This will neutralise—and render non-hazardous, any biological agent within approx 5 minutes. However, this solution should not be used in the eyes, Abdominal cavities or on nerve tissues.

(b) ***Physical Method***: Physical methods are undertaken to render BW agents harmless through such physical means as heat and radiation to include auto airing. This method pertains to decontamination of equipment and protective clothing wherein they are subjected to dry heat at 160° C for 2 hours or with steam at 121°C for 20 minutes at 1 ATM over pressure.

## Chemical Agent Contamination

The process of decontamination of Chemical agents will be undertaken by following means:

(a) ***Initial Decontamination***: During initial decontamination at the decontamination centres, bandages are removed and wounds are flushed with plain water. The bandages are replaced only if bleeding occurs. Tourniquets are replaced with clean ones and the site of original, tourniquet decontaminated. The new dressings are removed in the operating room and submerged in 5% hypochlorite or placed in a plastic bag and sealed.

(b) ***Dry Decontamination***: All casualties on being brought to decontamination site are completely undressed. The contaminated cloths are placed in numbered plastic bags. Dry decontamination is carried out with Fullers earth (constituent of PDK II) or talcum powder, I flour in case Fullers earth is not available For at least 2 minutes, the complete body surface, except the eyes and exposed wounds, should be subjected to dry decontamination.

(c) ***Wet Decontamination***: The casualties after undergoing dry decontamination are subjected to wet econtamination. The wet decontamination is carried out by thorough scrubbing of the body surface with soap and 0.5% hypochloride or mild alkalis followed by a clean shower. The ambulatory and non-ambulatory casualties should undergo wet decontamination for minimum duration of 4 to 8 minutes respectively. It must be ensured that decontamination duration and quality are strictly adhered to, as even a drop of certain chemical agents may cause death.

## Evacuation

The aim of casualty evacuation in a phased manner from the site of disaster to the nearest medical echelon and thereafter, is to ensure that all casualties get necessary treatment as early as possible at the requisite place. However, in a WMD scenario the aspect of casualty evaluation gets badly degraded due to: -

(a) Massive number of casualties.

(b) All kinds of contamination viz. nuclear, chemical or biological.

(c) The necessity to decontaminate casualties prior to giving any form of treatment, resuscitation or surgery.

(d) Danger of contamination to medical and paramedical personnel in the chain of evacuation.

In addition, other factors likely to effect speedy and efficient casualty evaluation are: -

(a) Shortage of manpower for handling large number of casualties.

(b) Need for a large number of ambulance vehicles for casualty evacuation separately for contaminated and uncontaminated casualties.

(c) The requirement of vast resources for items like casualty bags, stretchers, medical implements and medicines especially for treatment of burns cases.

## Modification of Hospital Infrastructure to Cater for Handling NBC Casualties

As has been stated earlier, the treatment of NBC casualties follows the conventional method and standard treatment protocols except for one important issue that of decontamination. To cater for the management of NBC casualties and to address the vital issue of decontamination of casualties, the hospital primarily has to be demarcated into two areas namely, clean area and contaminated area by a prominent line or marker. The closed covered area of the hospital could be designated as the clean area and any open area in front of the casualty department if, adequate or any other suitable area selected by the hospital administrative areas could be earmarked as contaminated area.

***Contaminated area***: This area should be demarcated prominently from the clean area as it helps in segregating the contaminated persons and prevents secondary contamination of hospital personnel and equipment. Casualties should not be permitted to cross the demarcation line. The contaminated area will be divided into two areas as under:

(a) ***Decontamination site***: Casualties on being brought to the hospital should be brought to the decontamination site. The decontamination site should preferably be located downwind of the hospital, so that the NBC contaminants move away from the hospital. Within the limiting constraints the endeavour should be to site the decontamination area in such a way that it will be possible to relocate within one km inside the hospital area in case there is an appreciable change in the wind direction. Medical and paramedical staff at the decontamination site would be required to be in complete protective gear and be suitably equipped with chemical and radiological detection equipment to monitor the efficacy of the decontamination procedure. The procedure of decontamination is both labour and resource intensive. It is estimated that each ambulatory patient would require two persons for effectively carrying out decontamination and will take approximately 10 minutes. Lying cases will require a team of four persons for carrying out the decontamination drill, which in the hands of trained personnel will take about 12-13 minutes per casualty. Initial triage will be carried out at the decontamination site and casualty suitably delineated by means of colour discs. Notwithstanding the fact that the priortisation of casualties would be done at the triage site by the Triage Officer / Medical Officer based on the clinical condition of the patient, the broad parameters / guidelines for classification of casualties can be grouped as under: -

| | | **Severity** | **Colour Code** |
|---|---|---|---|
| (i) | Ambulatory Patients | Mild Casualty | Green |
| (ii) | Non- ambulatory Patients<br>(aa) If breathing spontaneously | Moderate Casualty | Yellow |
| | (ab)If requiring mechanical ventilation | Severe Casualty | Red |
| | (ac) Moderate to severe conventional injuries associated with contaminated wounds | Combined Casualties | White |

(b) ***Regulation Site***: It is opened next to the decontamination site and is brought into use only when the decontamination site is overloaded and unable to accept any further casualties. Casualties who are brought here will only be administered initial treatment till they can be transferred to the decontamination site. No efforts at decontamination will be undertaken at this site. Staff at regulation site should be in protective gear.

**Clean Area**: The clean area is demarcated from the contaminated area by a prominent yellow line. The implications for demarcation of a clean area are as under:

(a) The medical and paramedical staff are not required to be in protective gear making it that much easier for them to handle the casualties.

(b) The casualties are completely undressed. The non-wearing of protective gear facilitates the medical staff in thorough examination and effective medical management.

**Demarcation of the Hospital Area:** In view of the different requirements of treatment for the different categories, it is recommended that the hospital be demarcated into four different areas as under:

**(a)** ***Mild Casualty Site***: Any conventional acute ward of the hospital adequately staffed and stocked would suffice for being designated as a mild casualty site. However, sources and measures for indenting chemical agent, specific antidotes should be known to the staff.

**(b)** ***Moderate Casualty Site***: Provisioning of chemical agent specific antidotes has to be catered for in addition to the facilities available in a conventional acute ward of the hospital.

**(c)** ***Severe Casualty Site***: Enhanced staffing and provisioning of adequate life saving and monitoring systems would be required. Preferably the ICU of the hospital could be designated as the severe casualty site.

**(d)** ***Combined Casualty Site***: Combined casualties i.e. moderate to severe conventional injuries associated with contaminated wounds may have to be taken up for emergency management without undertaking the decontamination drill. This implies that these casualties will be allowed to cross the yellow line without undergoing decontamination and will constitute a contaminationhazard inside the clean area. To cater for this contingency it is imperative that the combined casualty site be located preferably in a different building, vis-à-vis the other three casualty sites, to lessen the chances of breakthrough of sanctity of the clean area. The staff deployed in the combined casualty site is required to wear double surgical gloves and facemask to ensure physical protection and avoid secondary contamination from now contaminated wounds. The medical and paramedical staff manning the various casualty sites is to be adequately briefed beforehand on the specifications of confirmed treatment protocol against various chemical warfare agents.

To facilitate standardised treatment for the above-mentioned different categories, the casualties immediately on arrival at clean area should be transferred to the different sites according to the severity of their symptoms (along with their medical case file). The paths to the different sites should also be designated by the colour code given to the different categories of casualties.

## Miscellaneous Facilities

***Psychological Counselling Site:*** This has to be manned by psychiatrists and counsellors to manage patients under extremes of emotional stress as a direct consequence of being exposed to NBC Warfare agents.

***Public Information Centre:*** The present concept of enquiry counters at hospitals can be re-modelled to function as Public Information Centre and be manned by professionally trained staff who are trained to cope with the stress conditions of family members and dependants of the casualties. Another modification, which could be considered for implementation is for taking photographs of unconscious patients on arrival at the clean site. One copy of the same can be kept at the Public Information Centre for facilitating identification by relatives / dependants.

***Movement Control:*** The most prominent short-term effect of biological or chemical weapons is the large number of casualities that they can cause. As a corollary to this aspect it is quite probable that the hospital nearest to the site of the incident will be engulfed by hordes of casualties reporting. In addition to this mass casualty scenario will be the added burden of contamination avoidance, which will require provisioning of a "clear route" and "contaminated route". To enforce these provisions and ensure strict traffic control, the aspect of movement control gains added importance. Officials responsible for ensuring movement control have to take care of the above aspects while formulating an effective and co-ordinated movement control plan, which should be practiced regularly.

## Measures Recommended To Combat WMD Incidents

In no part of the world, the public health and medical authorities are adequately geared up to detect and respond to bio and chemical hazards/disasters, natural or inflicted by terrorists. Some aspects of preparedness against WMD incidents are similar to those needed to face natural disasters. Some urgent measures, which are recommended for implementation, are:

(a) ***Stocking Vaccines and Drugs***: We need very large stocks of vaccines and drugs including specific antidotes against chemical warfare agents in time of disaster. Care should be taken that the drug manufactures do not exploit the situation by hiking the prices. Where a particular need for equipment, antidotes and antibiotics or vaccines has been identified, pre-attack stockpiling and planning of distribution system or designation of sources of rapid supply to make them available to the exposed population will be necessary.

(b) ***Diagnostic Facilities***: Because early diagnosis of both biological and chemical exposure will be important in the choice of treatment and responses, preparations should include the establishment of a reference laboratory (or a network of laboratories in large areas) in which potential agents can be identified. In addition to the need for diagnosis for purpose of medical treatment, samples obtained from a delivery system or the environment or from patients, will require forensic analysis. Early diagnosis will be facilitated if regional laboratories have the necessary equipment and staff for that purpose. In our context the premier laboratories i.e. National Institute of Communicable Diseases (NICD), School of Tropical Diseases, DRDE, AFMC etc. should have BSL-III/IV capabilities with state of the art communication facilities and equipment.

(c) ***Disease Surveillance System***: Routine sensitive and near-real-time disease surveillance systems are essential in both disease outbreaks and those caused by biological agents. Such systems should be in place well in advance of an attack, so that the background disease prevalence in the area concerned is known. A national centre can detect a national outbreak not noticed in any individual region and it can also economically provide epidemiological expertise for investigating the cases and the source of outbreaks. Establishment of mechanism for the routine exchange of information between the public health and veterinary sector is also very important, as many biological agents are zoonotic.

(d) ***Contingency Plans Of Action***: Authorities should draw contingency plans of action separately for each vulnerable area, to meet the situation when it develops. Hospitals in the high threat areas should be identified and basic infrastructure modification to meet the requirements for handling chemical agent casualties like portable open stay shown, earmarking contamination situation measures for safe effluent disposal etc. should be undertaken.

To conclude, ideal management of casualties in NBC scenario can be based on the experiences gained from natural disasters and conventional war. Medical management in WMD terrorist incident/disaster poses many challenges. The medical services will have to gear themselves up to face the problem like large number of casualties occurring practically instantaneously, lack of experience in dealing with such types of situations and problems of combined injury syndrome. There should be a rapid and effective triage at the site of the disaster, a sorting and classification of casualties according to severity of the injury and the emergency treatment required. An organisation of medical care with adequate trained staff, good communications , ample medical supplies, good transportation facilities and not too many casualties, can function effectively.

# 10

# Role of Media

*"The media are the terrorists' best friend. The terrorists' act by itself is nothing".*

**Walter Laqueur**

*'I know a guy never mind his name – who was on one of those government terrorism commissions – never mind which one and used to say I ought to talk to him. I never did. I was too busy, not just with Bill and Monica but with other things as well, some of them very important. Anyway, I never wrote about the terrorist threat to this country. I was negligent'.*

**Richard Cohen**
**A Columnist in United States**

Media has an important role to play while managing the crisis and consequences management stages of an WMD attack. It is an important tool, which is exploited both by the terrorists and the state, but for different purposes. During the 11, September 2001, terror attacks and the subsequent anthrax threat in United States, as public officials struggled to respond and hunt for the perpetrators, the media proved indispensable. Television, telephones, radio, newspapers, news magazines and the Internet amounted to a communications network that, at its best, created virtual public meeting places and at its worst, exploited highlight replays of human evils most horrific images of destruction.

Mass media plays a central role in the terrorists' scheme of things. The perpetrators of 11 September attacks have exploited the media more shrewdly than other terrorists, but their deeds have affirmed the media's centrality in the calculus of terrorism, regardless of whether the violent deeds are major or minor in size. Historically, terrorists vied for publicity (having to settle for mouth-to-mouth reporting before the invention of the printing press), now they can exploit far-reaching, instant and global media networks and information highways to carry news of their violence along with what has been called "propaganda of the deed".[1] Indeed, as the weeks following the attacks on the World Trade Center and the Pentagon and the US counter attacks on targets in Afghanistan made utterly clear, terrorist such as Osama bin Laden and his terrorist organisation — the al Qaeda — can be formidable players against powerful nations – even the United States, the world's only superpower – in a propaganda campaign triggered by violence.

It is well recognised that some form of symbiotic relationship exists between the news media and perpetrators of spectacular terrorist incidents.[2] One of the most important aim of a terrorist attack is to gain publicity for a particular cause. In some cases, publicity is the sole aim. The presumed primary aim of the media is to inform. Competition between media organisations seems to heighten the necessity to focus on the emotion generating as opposed to purely informational aspects of news reporting. Terrorists are well aware of this phenomena and consciously script what is termed 'live- action spectaculars', news events that cannot be ignored by the media.[3]

The media in any democracy cannot ignore an event that has been fashioned specially for their needs. Terrorists can no more do without the media than the media can resist the terror event. The two are in a symbiotic relationship, so that any restriction of one narrows the bounds of the other. Noted terrorism expert Brian Jenkins argues that by definition, "terrorism is a product of freedom, particularly, of freedom of the press".[4] Indeed the correlation between the escalation of global terror-violence in the past twenty years and the innovations in media technology that facilitate rapid dissemination of information to large audiences is more than mere coincidence.[5] Of all the foundations of a free democratic society, the most basic, the freedom to know, to be informed, has guaranteed that such knowledge and such information can be fashioned by the fanatic through the conduit of the media eye. To close that eye would erode that fundamental right and would close an open society. Yet, not to do so assures future massacres, further terrorist events with little hope of audience saturation.

The terrorists' deliberate goal is to acquire publicity in the form of media coverage. In times of major natural and manmade catastrophes, the public depends on the news, especially television and radio reports, for information about the crisis at hand and for instructions about what to do and what not to do.

When terrorists fail to explicitly claim responsibility for acts of violence, the mass media - simply by reporting extensively on such incidents - transmit the perpetrators' messages by warning citizens that even the most powerful government cannot protect them from this sort of violence.

According to the Manual of the Afghan Jihad that was used as a training guide for bin Laden's al-Qaeda terrorists, publicity was always the overriding objective. Thus, the Manual recommends targeting "sentimental landmarks", such as the Statue of Liberty in New York, Big Ben in London and the Eiffel Tower in Paris because their destruction would generate intense publicity with minimal casualties.[6] If at all possible, terrorists just like actors who work in the non-violent, legitimate political process - prefer to circumvent the media gatekeepers and express their messages directly in front of their target audiences. In the Mini manual of the Urban Guerrilla, Carlos Marighela wrote about a two-fold approach to the terrorists propaganda scheme and suggested that the modern mass media, simply by announcing what the revolutionaries are doing, are important instruments of propaganda. However, their existence does not dispense fighters from setting up their own secret presses and having their own copying machines. The war of nerves or the psychological war- is a fighting technique based on the direct or indirect use of the mass media.[7] Marighela recommends that revolutionaries should tape messages and acquire radio stations to broadcast them; some contemporary terrorists groups own their own radio transmitters and/or television stations. Thus, the Revolutionary Armed Forces of Colombia (FARC), the longest surviving Marxist, so called guerrilla, army that controls large parts of the country, fight over the airwaves by broadcasting its "Voice of Resistance" with their six mobile transmitters, and the Hizbollah organisation broadcasts over its own television stations in Lebanon.[8]

On the other hand, in case of violence committed by states, governments are not at all interested in publicising domestic or international acts of violence that harm innocents and non-combatants. In sharp contrast to terrorists, governments, especially democracies, will speak of civilian casualties as the unintended victims and of damage caused to innocent bystanders as regrettable consequences of war in the hope of less, rather than more, publicity. Moreover, when governments commit or sanction domestic political violence in their own countries or abroad, they and those who act with their support do not want to publicise their deeds to larger audience. As usual, one finds exceptions to the rule. In this context, exceptions apply to governments involved in what they call anti-terrorism (preventive measures) or counter-terrorism (reactive measures) or the use of force against persons accused of perpetrating political violence. In such cases, publicity may be considered helpful in a government's efforts to discourage further political violence and to satisfy a fearful domestic public that demands protection. In late 2000, for example, Israeli officials acknowledged " publicly, explicitly and even proudly that their military was targeting and killing individual Palestinians accused of actual attacks on Israeli citizens or of planning such deeds". While Palestinians spoke of "state terrorism" and "assassinations", a highly placed government official in Israel defended those acts as counter-measures and said that the "most effective and just way to deal with terror is the elimination or incarceration of the people who lead these terrorist organisations".[9]

The inevitable and primary role of communication and propaganda in the terrorists' design and the contemporary mass media's appetite to facilitate the need of virtually all terrorists to have their deeds publicised. Former British Prime Minister Margaret Thatcher had it right when she proclaimed that publicity is the oxygen of terrorism. If anything has changed in the last ten or fifteen years, it is the increased availability of the sort of oxygen Mrs. Thatcher warned of and upon which terrorism thrives.

Over the years, the mass media has changed in dramatic ways - mostly, but not solely, because of the global reach of the internet and cellular phones. In the last decade or so, when more and more people are connected to and communicate through the Internet, the traditional and old media transformed too, firstly by establishing their presence on the World Wide Web and secondly, by accelerating mergers and acquisitions that created ever larger mega media organisation with national, international and even global character.[10] Today, the vastly expanded public spaces provided by the old and new media offer groups that promote and commit political violence far greater opportunities to communicate with their various audiences than in the past. Moreover, the internet has become a target and a means to inflict damages on governments, businesses and private citizens. Cyber terrorism or internet-terrorism are real threats to computer dependent politics and economics. Indeed, it has been argued that virtual attacks represent the new warfare of the twenty-first century, but, by the same token, information attacks are the new terrorism of our time.[11]

As serious news organisations move increasingly away from reporting what journalists – gatekeepers - deem important for the enlightenment of fellow citizens to what profit-oriented corporate managers consider interesting for the entertainment of news consumers, "hard' news is increasingly crowded out by "soft" news.[12] As a result, most news offered is essentially a blend between "hard" information and "soft" entertainment – infotainment in the guise of news reporting. The round-the clock, " all the news all the time," television channels and the proliferation of the news magazine format on television has driven this trend. Infotainment, far more than informative hard news, strives on the very images and themes that terrorists incidents offer - drama, tragedy, shock, anger, grief, fear, panic – the ideal ingredients for transforming real life terror into breathtaking thriller or heart breaking soap operas designed to captivate and stir up audience. Journalists are aware of the impact that powerful visual images of terror have on those who are exposed to them. An example of this is the bomb attacks carried out by the terrorists in Philippines in December 2000 that killed twenty-two people and injured 124 men, women and children. The Associated Press reported from Manila "graphic media images heightened the sense of insecurity". The television showed an unconscious boy in the hospital after his leg was amputated, and a newspaper carried a front-page photo of a rescue worker carrying a little girl's mangled body.[13] Whether the deadly bombs were detonated by members of one of the country's Muslim separatist groups, the Moro Islamic Liberation Front or the Abu Sayyaf, or by the left-wing New People's Army mattered less than the fact that the grim media images in the wake of the multiple terror action shook up the populace, the government and the law enforcement community - certainly a major goal of the perpetrators, who wanted to undermine the Philippine people's trust in their government's ability to protect them.

## Internet

Terrorists who fight modernity and its perceived evils in the name of defending traditional values and religious principles do not shy away from enlisting advanced technology for their holy wars to enforce their agendas. The same is true for governments and movements that encourage and support terrorism. Osama bin Laden and his associates, for example, appear traditional in the way they dress, speak, live and preach. But, like similarly inspired groups, they rely heavily on members in their midst and network of compatible groups and individuals that are trained in and familiar with the predominant communication technologies of air time. Thus, operatives in al-Qaeda and other terrorist organisations are known to be computer literate; they compose training manuals on their word processors, distribute them on CD-ROMs, or transmit their files via e-mail to trusted comrades. Bin Laden and members of his group were in a habit of plotting and discussing acts of terror by e-mail, but on learning that e-mail exchanges are as susceptible to intercepts as are satellite telephone transmissions, they circumvented the most advanced electronic surveillance techniques by resorting to encryption in the form of software technology that locks computerised information so that it cannot be accessed by unauthorised persons and organisations.[14] In principle, the purpose of these encryption programs - the protection of people's privacy - was and still is laudable. But, the very same technology has made it difficult for intelligence and law enforcement agencies to pick the electronic locks of information flows circulating among terrorists.

## Consequence Management and Media

In any kind of emergency situation, whether natural or man made, media is the first to provide details of the incident. In today's world, it has become evident that even top government officials rely more on media information than their own sources, for the obvious reasons of the television channels being the fastest to report and show live images of the incident. The response professionals on site are the most authoritative sources of information, and thus they are most sought out by the media. This is an ideal position from which to convey important information to the public at large and to communicate with the victims, with victims' families, friends and neighbours, with the perpetrators and their supporters, and, last but not least, with political decision makers in various jurisdictions. In short, under these circumstances, those who respond to a terrorist event have the opportunity to manage information and influence the public in what they believe is the best solution for resolving an emergency situation of this kind and/or dealing with the aftermath of a WMD incident.

Terrorism is a means to communicate messages to the public and to the governmental decision makers. The terrorists on the one hand and crisis managers and response professionals on the other, compete for media attention. The mass media is infatuated with terrorism and have modest interest in counter-terrorism. This disparity is even more extreme with respect to anti-terrorism. The act of terrorism is a master key for unlocking the door that grants access to mass media. This means that crisis managers and response specialists compete with the perpetrators of political violence in that each side wants to have the loudest and most persuasive voice and message. In this competition, terrorists seem to start with an advantage because their violent deeds are a powerful message that commands the mass media's attention and, thus, of their target audience(s). But, response specialists and crisis managers (such as members of police and fire department, emergency medical teams etc). and political leaders (such as chief ministers, mayors, bureaucrats etc) are in excellent positions to dominate the news because they are part of one of the cornerstones in the "Triangle of Political Communication" with formal and informal links and relationships in place before emergencies arise.[15] Finally, victims of terrorism, their families and friends and the public at large may also have their own ideas about how a particular incident should be handled that are different from the response specialists' at the scene and the crisis managers' in their command centres. This is likely to be the case during hostage situations, when the authorities may not want to give in to the demands of hostage-keepers, while hostages and their loved ones may press for release of captives at any price. It is clear then that the interests of mass media, terrorists, their victims, all kinds of societal groups (such as the relatives and friends of victims) as well as response specialists call for a delicate management of public information and media relations on the part of those who respond to and deal with terrorist crisis. News media and terrorism response specialists have very different objectives: while where the mass media aims to "scoop" their rivals with news stories that will grip and sustain the public's attention and hence, increase their ratings and revenue, the police, on the other hand, are first and foremost concerned with the protection of life, the enforcement of law and apprehending those guilty of committing the terrorist violence. There have been numerous examples where the efforts of the police have been directly threatened by the behavior of sections of the media. The mass media can and does interfere with the plans and intentions of those who are in charge of handling terrorist incidents/crisis. But the same reporters and news organisation can also be very helpful and in fact, crucial in efforts to manage such incidents. While each terrorist crisis has its own complex characteristics and therefore, calls for particular response, there are some basic rules, caveats, experiences and practical examples to draw from.

Keeping the constraints and the need of media to ward off of challenge in today's operating environment, the terrorism response professionals and crisis managers need to follow certain rules and guidelines. The response mangers are required to provide the media with steady flow of information so that they can meet their deadlines otherwise they will lose their advantages as the most authoritative news source. If denied information it will turn to those who are closest to the emergency, criticise the way the crisis is handled resulting in rumours and not facts, being reported. If, the crisis managers give information to the media representatives they maintain their vantage point from which to frame and shape crisis information and construct the predominant story line. Today the press has become far more competitive and aggressive in a media landscape that is increasingly shaped by a large number of all-news radio and television channels that thrive on breaking news stories with live broadcasts. As a result, not only small town police personnel and other emergency respondents but also their colleagues in far larger jurisdiction are challenged by an onslaught of media during a crisis.

Media, especially television and radio, are the most effective means to reassure and calm an unsettled public. When terrorists strike, they kill, hurt and frighten their immediate victims in order to get the attention of and influence the general public and government decision makers. As they strike innocent persons and demonstrate the impotence of the targeted government to protect its citizens, terrorists spread fear and anxiety in their target audiences and could even cause panic in the threatened society. This is likely to lead to political instability and make citizens wonder whether their government and its officials are fit to prevent terrorism in the first place or at least effectively respond to acts of political violence. For these reasons, crisis managers must utilise news media in order to project the image of professional, resolute and competent leadership even in most difficult of circumstances. Although, crisis managers and response professionals probably have more pressing problems than appearing before the media, they still need to find some way to disseminate information to the news media if only to let the populace know what is being done, that political leaders and the response community are dealing with the crisis in the best manner they can and that the things are under control. The crisis managers, by appearing in news conferences and granting interviews can best exploit the extraordinary public attention in order to get their message through, to demonstrate their composure, and to convince citizens to keep their cool and trust that the authorities will do everything possible to protect them.

The emergency response community must utilise the mass media as a vital instrument for enhancing and coordinating their actual emergency efforts by telling the public what to do and what not to do. In times of major emergencies, the radio and television become vital arms of the government because these offer the crisis managers unlimited access to communicate with the public, with emergency specialists and with other government authorities. Through this medium, the officials can convey their needs, concerns and warnings indirectly through reporters and other media personnel. This was amply demonstrated during the Gujarat Earthquake in India and the 26 December 2004 Tsunami Disaster in Asia, where the medium of television, radio and press played its part by informing the audience of the magnitude of the disasters. With this exposure the gravity of the situation was understood and aid poured in from all over the world. The donors were informed of the materials needed and also, where the same could be delivered.

Trained personnel must monitor the mass media round the clock for accuracy in crisis related news and during hostage situations, for news items that could endanger the lives of the hostages and law enforcement officers and hamper efforts to end a standoff. The response community must prepare for reliable personnel inside emergency command centres to monitor mass media throughout the critical phases of a crisis in order to ensure that publicised information is correct and that appeals for assistance of one kind or another are made only when prepared and backed up by organisational moves to facilitate responding citizens;

for example, a vast number of blood donors ascending on hospitals or schools or emergency workers coming from afar to assist in rescue tasks or people massing medical facilities to receive antidotes for released biological or chemical agents. To do this job well, monitors must be close to crisis managers. If published information is wrong or contradictory, if rumours are elevated to the level of hard news, if appeals to help have resulted in insufficient offers, if a stricken population is misdirected with respect to treatment facilities, then the media must be contacted immediately so that mistakes are corrected or changing circumstances are reflected in the news. Given the all-out competition between news organisations, the pressure to present breaking news, the determination to report some new angle although a terrorist situation has not changed and the tendency to sensationalise even genuinely dramatic situations, the hastily reported and often unverified news is likely to contain inaccuracies, mistakes and problematic features. The underlying importance of monitoring news by crisis command centres and reacting promptly to exploit new opportunities for crisis resolution, law enforcement etc. or to prevent potentially harmful consequences of publicised information cannot be overemphasised.

Response specialists must keep in mind that most journalists and news organisations cooperate with the authorities faced with truly serious incidents. In this cooperative mode, the media can be instrumental in resolving these types of crises. While reporters are chasing exclusives in order to scoop the competition, many of them are not inclined to disregard that they could endanger the victims of terrorism especially hostages and the members of rescue teams by reporting indiscriminately and revealing tactical information. In some situations, law enforcement officials and other response professionals may consider asking media organisations for assistance and cooperation.

Although representatives of the news media will demand access to the site of a terrorist act, response personnel have the right and the responsibility to limit or deny access if the presence of media representatives threatens the safety of victims and response personnel or inhibits rescue efforts, negotiations or other means to resolve the emergency. The point here is not to encourage law enforcement officials and other response professionals to automatically deny media representatives access to incident sites, whenever terrorism occurs. It should be clear from the previous guidelines that such a policy and practice would backfire because it would surely create an advers relationship with the fourth estate, risk the response community's status as primary news source, interfere with crisis managers need to inform the public and diminish chances of their cooperation with the media. But, when the presence of media interferes with the work of rescuers, negotiators or other response professionals, law enforcement officers have the right to limit or deny media access. Police authorities, as a matter of course, should develop clear guidelines governing the media's access to the scene of a terrorist incident. The media should be made aware of these guidelines and conditions before terrorist incidents and similar events occur. This step seeks to avoid the arguments and recriminations that can develop between individual reporters and police authorities during the rush and confusion of violent incidents.

As a general rule, crisis managers should not go on the prior restraint route. In almost all terrorist emergencies, curbing press freedom is not an option. However, in extraordinary cases, when information–if revealed - would result in a clear danger for human life or the national interests in very specific and most serious ways, response professionals must sometimes consider this extreme and highly controversial step. This is true in a hostage rescue scenario, where the Special Forces are planning and preparing to attempt special operations to free the hostages. Terrorists holding the hostages would be following the events in minute detail to know what is happening by watching TV and monitoring radio broadcasts. In such a scenario, it is for the media to exercise restraint and not give any hint of the impending mission. This act of theirs would save precious lives of both the hostages and the Special Forces personnel.

An example of such an incident in the new millennium is the case of Abu Sayyaf terrorist group demanding independence from the Philippines. The group held hostages for many months and on several occasions, threatened numerous times that rescue efforts by the Philippino Army would cause the execution of hostages and actualised these threats on several occasions. On the other hand, if the security agencies learn of an imminent terrorist strike with a radiological, chemical or a biological attack and while planning to prevent terrorists from carrying out the attack, the news leaks out, through media it is not far fetched to assume that the news of an imminent attack with weapons of mass destruction is likely to result in hysteria; panic, unorganised mass flight or loss of life, among the population. Here, a case could be made for prior restraint and preventing the alarming news from being publicised.

However, it would be prudent that only in extreme situations, when there is no doubt that the pending publication of a certain information will lead to irreparable harm most likely to hostages, law enforcement authorities and other response professionals - is the prior restraint route a realistic option for those who manage and respond to terrorist acts.

Encouraging media guidelines for reporting terrorist incidents may be prudent; however, trusting that news organisations follow these guidelines is not sure. Neither formal guidelines nor informal newsroom agreements on how to handle the coverage of terrorist incidents have eradicated excessive and potentially harmful reporting. If anything, the growing competition in the media market and the trend from hard news to infotainment news has led to more aggressiveness in pursuing the sensational, dramatic, tragic and frightening aspects of news in general and in the area of terrorism incidents in particular. Even in news organisations with the best intentions to adhere to self–imposed guidelines, these codes of behaviour and reporting fly out of the window in the face of major incidents, when competitive juices flow especially energetic in the struggle for ratings, circulations and most of all, corporate profit imperatives.

It certainly makes sense to encourage news organisations to consider and discuss how to cover, or not to cover, terrorist situations regardless of whether the results are formal guidelines or simply better understanding of the problems surrounding this sort of coverage. Over time, such practices and measures could affect terrorism coverage for the better. But, unless all news organisations sign on to what critics will call self-censorship the competitive tendencies will prevail and short-circuit meaningful changes.

By showing their human face in dealing with terrorism's victims and their loved ones, members of the response community will help others and themselves. When terrorists strike, response specialists must deal with victims as well as with their families, friends, neighbours and the larger public on one level, crisis managers and everyone else involved in dealing with an act of terror are expected to be cool, rational, detached enough to effectively handle whatever has to be done. On another level, however, the same people are expected to be sensitive to the feelings of the immediately affected people and responsive to their needs. In no other area of emergency response, feelings of both gratitude and resentment more easily and more vehemently expressed than in the real or perceived treatment of victims and their families. The visuals of exhausted emergency workers who helped the victim of a terrorist strike from the scene of unspeakable terror, the sensible words of officials speaking for the police and fire department and other organisations added up to a composite crisis response community with a human face that was comforting to the injured, to the families of the victims and to the nation. The unspoken message here was that of men and women who went beyond the call of duty in their difficult work and were at all times a part of a community in shock and pain.

All organisations, public and private, need to realise that terrorism, like other crises, calls for both professional and human response. To show compassion by deed for the victims of terrorism and their loved ones is certainly the most important imperative here, but to translate this reality into a media reality is another. Certainly, the victims of terrorism and/or their loved ones will be thankful when they encounter compassionate response professionals. But beyond this, it is also important that crisis managers and those who speak for the emergency response community project this compassionate and comforting image via the media to a traumatised public. Public perception is not shaped by reality but by the pseudo reality reflected in the mass media.

The crisis managers and terrorism response professionals can circumvent the traditional news media and communicate directly with each other and more importantly, with the public by using the Internet. Terrorism response professional can by-pass the media and directly communicate with the public through the medium of Internet. In this manner those personnel will be most successful in getting their own messages across the public if they utilise their websites, which are most authoritative emergency information. Such Internet sites need to be updated often and diligently and are additional means to provide new press releases, statements, warnings, clarification etc. Internet also provides an ideal means for response professionals to communicate with each other, especially when geographically apart. Internet can, therefore, serve as a more specialised means of communications between various elements of a response to a terrorist incident by offering a shorter and quick communication, action and cooperation.

The media has a responsibility to fully inform citizens of important events and developments. Government's power on the other hand, especially its authority in extraordinary situations needs monitoring and robust checks to prevent abuse of power. If there are abuses, media must ensure that are revealed. But, the media must remember their responsibility to the public interest and public good. A responsible press must not only insist on factual and full information but must also recognise in extra ordinary cases that certain revelations in the news could have devastating consequences for rescue personnel and the victims of terrorism.

Publicity, they say, is as essential for terrorists as the air they breathe and, therefore, they will continue as in the past with horrific acts of violence for a great deal of public attention without spending large amounts of money. Terrorists perform their violent acts not only to harm victims but also to obtain the attention of masses that they could not reach without the media communicating their deeds and carrying their messages. Inadvertently, then, the media plays a central role in the calculus of political violence and are put into positions where they can magnify or minimise these kinds of acts and their perpetrators, or, of course they can provide coverage that avoids either one of those extremes. On the other hand, media plays a very important role in effectively managing the consequences of WMD attack by terrorists by projecting the true picture of the damage that has taken place and also by guiding the response personnel and the relief material to the places where it is required the most. Media can also play an important part in educating and preparing the population to face such crisis.

## References

1. Schmid, Alex P. and Jenny de Graaf, *Violence as Communication: Insurgent Terrorism and the Western News Media.* (Beverly Hills, Calif. Sage Publications, 1982), pp. 11-14.
2. Wardlaw, Grant.*Political Terrorism: Theory, Tactics and Countermeasures,.* (Cambridge University Press, London, 1989). pp.74-81.
3. Laqueur, Walter. *The Futility of Terrorism,* (Harpers, 1976). pp 102-107.
4. Jenkins, Brian. *Research in Terrorism: Areas of Consensus, Areas of Ignorance.* (Washington DC, 1995). Pp.23-32.
5. Lawerence Zelic Freedman and Yonah Alexander, *Perspectives on Terrorism.* (Hindustan Publishing Corporation, Delhi, 1985). Pp.177-185.
6. Hamza Hendawi, *"Terror Manual Advises on Targets"* http://story.news. yahoo.com/news?tmpl=story&u=/ap/20.../afhan_spreading_terror_(accessed 11 Sep 2004).
7. Schmid and de Graaf, n. 1, p.19.
8. Sontag, Debroh. *"Israel Acknowledges Hunting Down Arab Militants". New York Times,* 22 December2000, p. A12.
9. Alger, Dean. Mega Media: *How Giant Corporations Dominate Mass Media, Distort Competition and Endanger Democracy.* (Rowman & Littlefield, 1998) P.126.
10. Adams, James. *"Virtual Defense".* Foreign Affairs 80. 3(May-June) 2001,pp. 98-112.

11. Bennett, Lance W. News: *The Politics of Illusion,* (Longman: New York, 2001) pp. 12-15.
12. *"Tracking bin Ladens' e-mail,"* Newsweek (21 August 2000): p.6.
13. *"Philippine police cite lead in hunt for bombers.* http://www.nytimes.com/aponlive/world/AP-Philippines-Explosions.Html. (Accessed on 15 Sep 2004)
14. Brigitte L Nacos, *Mass Mediated Terrorism,* (Rowman & Littlefield Publishers Inc, Oxford, 2002), p.169.
15. Ibid, p.169.

# 11

# Conclusion

The prospects of WMD terrorism and disaster have increased manifold because of the greater accessibility of expertise, material and easy availability of dual use technology. The likelihood has further been enhanced by the changing nature of international terrorism, with its more diversified range of motives, trans-national networks and capacity to mount coordinated attacks. A WMD terrorist attack/ disaster is a high consequence event. The best defence lies in its prevention but in the eventuality of a WMD attack, India must be prepared to employ all resources and retaliate effectively against any terrorist group or state launching such an attack. This capability and the political will to use it will act as an effective deterrent against a WMD attack.

There is a need to evolve a WMD disaster/terrorist incident strategy, so that relationships between various players likely to be involved in managing the event can be built, telecommunication infrastructure strengthened, medical surge capacity enhanced, laws passed to enable appropriate surveillance and prosecution methods and psychological care capabilities improved. All these challenges need to be confronted with full vigour. There are, however, considerations that distinguish a WMD terrorist attack from a natural or other man-made disaster. Some of the considerations that are unique to WMD represent relatively new issues for the public safety community.

There are a large number of agencies working towards the consequence management of WMD incident. A well established working relationship and willingness to prioritise cooperation over inter-agency or inter-government, competition is essential to planning for and responding to a WMD attack. There is a need to build up relationships that bridge inter-agency divides. This is an important aspect of preparedness, but it is too narrow of focus. District and state government's agencies must be included in planning, training and even routine meetings. Central law enforcement and other agencies must work closely with state and local police, public health, fire and EMS and recovery services since local agencies will be "on the ground" at the beginning and conclusion of any disaster. Communication between all levels of government will facilitate the building of relationships and alert the national government to the needs of local response agencies. Additionally, relationships must be established between military personnel and civilian first responders before disaster occurs. These two groups need to plan and practice interacting with one another through drills. This is the best way to ensure that, in a mass casualty attack the military has a clear role commensurate with the constitution and expectations of the Society.

The legal preparation is a vital but often overlooked aspect of domestic preparedness. Therefore, it is even more essential that laws be in place before an attack occurs. This enables the nation to determine, with appropriate rational debate, the limitations of investigations and other law enforcement powers. Additionally, robust legal preparedness affords law enforcement the necessary powers to investigate and prosecute those who possess or attempt to use dangerous but not yet illegal chemical, biological or nuclear components or weapons.

Although, evidence shows that people behave rationally in the face of disaster, there is heightened risk of immediate panic, short-term trauma, and long-term post traumatic stress disorder following a WMD attack. Therefore, it is important to tailor psychological intervention to the demands of the occasion. Contingency plans for WMD terrorism should be put in place so that counsellors are prepared to deal with particular traumas associated with an intentional use of weapons of mass destruction.

The local police fire and medical personnel will carry out most of the emergency response efforts, especially in the immediate aftermath of an attack. Assistance from the central level is necessary, but the central government cannot and will not be the dominant player in the first minutes or hours after an attack. Therefore, it is essential that emergency responders in state and local government be prepared to deal with an unannounced attack with a weapon of mass destruction. At the same time, national level coordination is vital to ensure that training and equipment are provided to localities in accordance with their needs and exercises are carried out that involve personnel from multiple agencies and levels of government. In addition, responders from all relevant agencies and all levels of government must practice skills in simulations that involve all aspects of response. Only in this way can these agencies, which may not be accustomed to working together, converge on the scene of a disaster and effectively carry out rescue and recovery operations.

## Recommendations

***WMD Detection Technologies:*** India should invest in the WMD detection technologies since, detection is the first line of defence and it is from where the tactical consequence management begins. This effort could be carried out by developing cutting edge technologies and also by working with friendly foreign countries.

***Response Assets*:** India at the central government level has got large standing army and central police organisations. There are a large number of state police forces. The need is to equip and train some of selected elements of these forces with specialised equipment at the centre and state levels to prepare them to deal with the crisis and consequence management stages of a WMD terrorist incident/ disaster.

***Awareness Training*:** New awareness, education and training concepts of consequence management are the keys to success.

These new concepts will ensure a lower common denominator of knowledge at the local, state and national level. This will facilitate the knowledge about the capabilities of other organisations. Towards this end, seminars and conferences should be planned at regular intervals. Talking about a problem is a first step, establishing awareness about the type of organisations and people involved is an another important step. Conferences offer significant national returns for a relatively small investment.

***Solicit Advice and Participation of Professional Organisation:*** The professional associations of police and fire departments, doctors, first responders etc. should be actively sought out. Their opinions will be invaluable in creating central standards that empower local response.

***Participants' Selection:*** The selection of participants in any aspect of consequence management should be carried out with great care. With so many organisations and agencies involved in consequence management, unity of command is probably impossible.o The only alternative is unity of effort based on however many people are needed at the table for collaboration and coordination. Once at that table, a transparent flow of information through all essential channels will make or break the cohesion of effort in response to a WMD attack. More than anything else this type of cohesion is the result of handpicked personnel with excellent interpersonal skills.

The threat of use of WMD by terrorists exists. The time to begin planning for managing the consequences is, now. There is a need to develop sound and effective strategies and implement the same both in letter and spirit.

Appendix A

# Biological Warfare Agents

**Characteristics of Selected Pathogens**

| Pathogen | Routes of Infection | Method of Dissemination | Untreated Mortality | Incubation Period | Treatment |
|---|---|---|---|---|---|
| Anthrax | S,D,R | Covert or aerosol 5% R-100% | S-less that | 1-6 days | Antibiotics (limited effect-ness' after symptoms develop). Vaccine available. |
| Cholera | . D | Cover or aerosol | 50% | 1-5 days | Antibiotics. Note: Oral rehydration lowers lethality. |
| Plague | V.R | Aerosol or Vectors | Bubonic-50% Pneumonic-50-90% | 2-10 days | Antibiotics, Vaccine available |
| Tularemia | V,S,R | Aerosols | 30-40% | 1-10 days | Antibiotics |
| Q Fever | V,R | Covert or Aerosol | Less than 1% | 2-14 days | Antibiotics |
| Venezuelan Equine Encephalitis (VEE) | V,R | Aerosol or Vectors | V-1% R-25-50% | 1-5 days | No specific medication. Vaccine under development. |

Contd.

| Pathogen | Routes of Infection | Method of Dissemination | Untreated Mortality | Incubation Period | Treatment |
|---|---|---|---|---|---|
| Viral Hemorrhagic Fevers (Ebola, Lassa, Rift Valley, Dengue) and Conjunctivitis | DC, uncertain | Aerosol | 40-90% | 4-21 days | Antibiotics |

*S-skin, D-digestive, R-respiratory, V-Vector, DC-direct contract.

Source: *JCS J-3 Weapons of Mass Destruction (WMD) Handbook,* U.S. Department of Defense, Washington, D.C., February 2001, p.37.

Characteristics of Selected Toxins

| Toxin Cytotoxin | Natural Source | Rate of Action | LD50 (MG/KG) | Effect |
|---|---|---|---|---|
| Microcystic (FDF) | Blue Green Algae | Rapid | 25-200 | Shivering, stupor, prostration, shock, liver enlargement |
| Ricin | Castor Bean or rapid | Can be delayed | 0.1-3.7; 1,000 (oral) | Nausea, vomiting, cramps, bloody nose, diarrhea, difficulty in breathing, twistching |
| Staphylococcus Enterotoxin Type B (SEB) | Staphylococcus aurous bacteria | Can be delayed or rapid | 0.023 (aerosol) | Vomiting, cramps, nausea, diarrhea, severe weakess |
| Tricothecene mycotoxin infected grain | Fusarium mold on | Rapid | 500 | Iching, tingling, vomiting, haemorrhaging, bloody diarrhea |
| **Mixed toxin types** | | | | |
| Snake Venoms/Toxins | Snakes | Rapid | 1-5,000 | Paralysis or haeomorrhaging |
| Cardiotoxins | Cobra, mamba, snakes | Rapid | 300 | Heart irregularities, lower blood pressure |
| Necrotic toxins | Pit viper | Delayed | 1000 | Haemorrhage,muscle destruction |

Source: *JCS J-3 Weapons of Mass Destruction (WMD) Handbook,* U.S. Department of Defense, Washington, D.C., February 2001, p.37.

Appendix B

# Classification of Biological Warfare Agents

Bio warfare agents comprise bacterial agents, viral agents and biological toxins. The more important and the most dangerous over along with the diseases caused are listed below. Categorisation of Bio Warfare Agents based on probability of use.

**(a) Bacteria**

| | | | |
|---|---|---|---|
| (i) | Bacillus Anthracis | - | Anthrax |
| (ii) | Brucella Abortus/ Melitenis/ Suis | - | Brucellosis |
| (iii) | Vibrio Cholera | - | Cholera |
| (iv) | Corynebacterium Diphtheriae | - | Diphtheria |
| (v) | Pseudomonas Mallei | - | Glanders |
| (vi) | Clostridium Perfringes | - | Glanders |
| (vii) | Pseudomonas | - | Meliodosis |
| (viii) | Yersinia Pestis | - | Plague |
| (ix) | Shigella Dysteneriae | - | Shigellosis |
| (x) | Francisella Tularensis | - | Tularemia |
| (xi) | Salmonella Typhi | - | Typhoid Fever |

**(b) Diseases Caused By Viruses**

(I) Ebola Haemorrhage fever

(ii) Eastern Equine Encephalitis

(iii) Congo Crimean Haemorrhagic Fever

(iv) Dengue Fever

(v) Hantaan

(vi) Lassa fever

(vii) Rift Valley Fever

(viii) SmallPox

(ix) Yellow Fever

**(c) Ricketssia**

(i) Ricketssia Prowazeki - Epidemic Typhus

(ii) Coxiella Burnetti - Q fever

(iii) Ricketssia Quintana - Trench Fever

(iv) Ricketssia Rickettsie - Rocky Mountain Spotted Fever

**(d) Toxins**

(i) Botulinum Toxin - Obtained from Clostridium botulinum bacteria

(ii) Saxitoxin - Cases paralytic shellfish poisoning.

(iii) Ricin - Obtained from Castor Bean

(iv) T-2 Mycotoxin - Obtained from Various Molds

(v) Aflatoxin - Obtained from various molds.

## Appendix C
# Chemical Agents and Their Effects

| Type of Agent | Symbol | Symptoms | Effect | Rate of Action |
|---|---|---|---|---|
| Nerve | GA<br>GB<br>GD<br>VX | Difficulty in breathing, sweating, drooling, convulsions, dimming of vision | Incapacitates at low concentrations Kills in sufficient dosage, VX is persistent and a contact hazard. The other agents are non-persistent and present an inhalation hazard. | Very rapid by inhalation, slower though the skin. |
| Blood | AC<br>CK | Rapid breathing, convulsions and coma | Kills in sufficient dosage. Non-persistent and an inhalation hazard. | Rapid |
| Blister | H<br>HD | No early symptoms | Blisters delayed hours to days; eyes and lungs affected more rapidly. | Blister delayed hours to days |
| | HL<br>L | Searing/ stinging of eyes and skin. | Immediate pain, delayed blisters. Persistent and a contact hazard. | Rapid |

| Type of Agent | Symbol | Symptoms | Effect | Rate of Action |
|---|---|---|---|---|
| Choking | CG<br>DP | Slightly lacrimatory | Damages and floods lungs. Death can result. Non-persistent and an inhalation hazard. | Immediate to 3 hrs depending on concentration |
| Incapa-citating | BZ | Rapid heart, vomiting, dry mouth, blurred vision, stupor, random action | Hallucinations, delusions, extreme excitement. Non-persistent and an inhalation hazard. | Delayed; 1 to 4 hrs depending on exposure. |

# Bibliography


## Primary Sources

1. Kargil Review Committee Report.
2. Reforming the National Security System: Recommendations of Group of Ministers (GOM), February 2001.
3. Feedback Report, Workshop On Weapons of Mass Destruction (WMD) Consequence Management at Washington DC, USA From 30 September To 02 October 2002.
4. High Powered Committee on Disaster Management Report
5. The 9/11 Commission Report – Final Report of the National Commission on Terrorist Attacks Upon the United States.
6. Minutes of interviews with a large number of serving and retired Officers of the three services, intellectual and academicians.

## Secondary Sources

*Books*

7. Alvin and Hiedi Toffler, *War and Anti-War: Survival at the Dawn of the 21 Century*, (London: Little Brown and Company, 1994)
8. Air Commodore Jasjit Singh, AVSM, VrC, VM, Kargil 1999-Pakistan's Fourth War for Kashmir, *Knowledge World*, New Delhi
9. Bruce Hoffman, "The Contrasting Ethical Foundations of Terrorism In the 1980s" *Terrorism and Political Violence*, Vol. No. 3 (July 1989), p.398.

10. Brain M. Jenkins, "Understanding the Link between Motives and Methods" in Terrorism With Chemical and Biological Weapons Calibrating Risks and Expenses, ed, Brad Roberts.

11. Bernard Stewart, ed. 'State-Sponsored Terrorism: The Threat and Possible Countermeasures, *Terrorism*, 8, No. 3 (1986), p.253-313.

12. Brian Jenkins, ed. International Conference on Terrorism and Low-Level Conflict (Santa Monica, California: RAND, 1982).

13. Jessica Stern, The *Ultimate Terrorist* (USA, First Harwood University Press, 2000).

14. Jeffrey D. Simon, Terrorists and the Potential Use of Biological Weapons, Report R-3771-AFMIC (Santa Monica, California: RAND, December 1989).

15. Maj Gen Arjun Ray, *Kashmir Diary* (New Delhi: Manas Publications, 1997).

16. Malcolm W. Browne, *The New Face of War*, The Trinity Press Worcester, and London, p. 965.

17. Nick Carter, *Disaster Management: A Disaster Manger's Handbook*, Asian Development Bank, Manila, Philippines.

18. Richard A. Falkenrath, Robert D Newman, and Bradley A. Thayer, America's Achilles Heel: *Nuclear, Biological, and Chemical Terrorism.*

19. Robert Kupperman and Jeff Kamen, *Final Warning: Averting Disaster in the New Age of Terrorism* (New York).

20. Richard A. Falkenrath, "Confronting Biological & Chemical Terrorism" *Survival* 40, No 3 (Autumn 1998), p.52.

22. Samuel Hutington, '*Clash of Civilisation and the Making of the World Order*' (New York: Touchstone 1997): See also, Samuel Hutington 'Clash of Civilisations,' *Foreign Affairs* summer 1993.

23. Walter Laqueur, *The New Terrorism: Fanaticism and the Arms of Mass Destruction* (Oxford University Press, 1999).

## Articles and Periodicals

24. Lawrence Freedman, "Revolution in Military Affairs," *Adelphi Papers*, 316 (London: IISS).

## NewsPaper Articles

25. Jeremy Lovely, " Hi Tech Bugs to Turn Future Battlefields into Mass of Censors," *The Asian Age* 21 May 2003.

# Index